64th Conference on Glass Problems

64th Conference on Glass Problems

A Collection of Papers Presented at the 64th Conference on Glass Problems

Waltraud M. Kriven
Editor

October 28–29, 2003
University of Illinois
at Urbana-Champaign

Published by
The American Ceramic Society
735 Ceramic Place
Westerville, OH 43081
www.ceramics.org

ISSN 0196-6219

ISBN 0-470-05146-9
ISBN 13: 978-0-470-05146-7

Contents

64th Conference on Glass Problems

Preface viii

Acknowledgements ix

Refractories

Surface Treatment of AZS Refractories using High Density Infrared Heating 3
T.N. Tiegs, F.C. Montgomery, D.C. Harper, C.A. Blue, M. Velez, M. Karakus, and R.E. Moore

Investigation of Defects in High Quality Glasses 13
K.R. Selkregg and A. Gupta

Review of Improved Silica Crown Refractory and Practices for Oxy-Fuel Fired Glass Melters 33
Alonso Gonzalez, John T. Brown, Rober P. Weilacher, Michael A. Nelson

Engineered FKS Platinum Solutions for High-Temperature Applications in Modern Glass Production 43
Michael Oechsle, Hubertus Gölitzer, and Rudolf Singer

Geopolymer Refractories for the Glass Manufacturing Industries 57
Waltraud M. Kriven, Jonathan Bell, and Matthew Gordon

Anomalous Thermomechanical Properties of Network Glass 81
John Kieffer, and Liping Huang

Energy and Combustion

Advanced Investigation Methods for the Characterization of Flames aimed at an Optimization of the Heat Transfer Processes in Glass Melting Furnaces 97
Axel Scherello

Alglass SUN: An Ultra Low NO_x Oxy Burner for Glass Furances with Adjustable Flame Length and Heat Transfer Profile 117
Bertrand Leroux, Nicolas Perrin, Pascal DuPerray, Patric Recourt, Rémi Tsiava, and George Todd

Glass Furnace Life Extension using Convective Glass Melting 129
Neil Simpson, Dick Marshall, and Tom Barrow

Fire Polishing with Premixing Technology 141
Hans Mahrenholtz

A Novel Furnace Concept Combining the Best of Oxy- and Air-Fuel Melting 153
Mark D'Agostoni, Russell J. Hewertson, Bryan C. Hoke, Jr., Julian L. Inskip, Michael E. Habel, Richard Huang, Kevin A. Lievre, and Aleksandar G. Slavejkov

How Mathematical Modeling Can Help to Reduce Energy Usage for Glass Melting 167
Erik Muijsenberg and Eric Trochta

Attenuaion and Breakage in the Continuous Glass Fiber Drawing Process 179
Simon Rekhson, Jim Leonard, and Phillip Sanger

Energy Conservation Opportunities in the Glass Industry . . .191
John D'Andrea

Process Control

Application of Fast Dynamic Process Simulation to Support Glass Furnace Operation 197
Olaf Op den Camp, Oscar Verheijen, and Sven-Roger Kahl

Application of Batch Blanket Monitoring System in Glass Furnaces 209
Jolanda Schagen, Ruud Beerkens, AnneJans Faber, Peter Hemmann, and Gunnar Hemmann

Thermal Imaging of All Furnace Internal Surfaces for Monitoring and Control 219
Serguei Zelepouga, David Rue, Ishwar Puri, Ping-Rey Jang, John Plodenic, and John Connors

Improvement in Glass Blister Quality by Throat Design 231
R.R. Thomas

Emerging Areas

Overview of the Activities of the Technical Committees of the International Commission on Glass 241
Henk de Waal

Recent Developments in Chemically Strengthened Glasses . . .253
David J. Green

Glass Art and Glass Science: A Mutually Beneficial Exchange 267
Margaret Rasmussen, Michael Greenman, and John Brown

Self-Repair of Glass and Polymers .281
Carolyn Dry

Preface

The 64th Conference on Glass Problems continues a long tradition, which was started on June 1, 1934, at the University of Illinois at Urbana-Champaign. Prof. C.W. Parmelee, the then head of the Department of Ceramic Engineering, initiated the first conference, which had an attendance of 50 men, representing eight states in the U.S. The first meeting was so successful that a second meeting was held on November 2nd of the same year, having an attendance of 97 men and 1 woman. On December 2, 1948, the Conference on Glass Problems was first held at The Ohio State University, and the two universities have been alternately hosting it each year since then.

C.W. Parmelee envisioned the mission of the conference to be "for the benefit of the glass manufacturers...especially in attracting the operating men for whom they were arranged." Today, the Conference on Glass Problems attracts both U.S. and international attendees. It has a high level of technical content and it provides a forum for learning and new ideas as well as information exchange. Attendees represent the glass manufacturing industries, suppliers, academia, and government laboratories.

On October 28-29, 2003, the 64th Conference on Glass Problems took place on the campus of the University of Illinois at Urbana-Champaign. The conference encompassed four topic sessions: Refractories chaired by Daryl E. Clendenen and Thomas Dankert; Energy and Combustion, chaired by Marilyn DeLong and Philip Ross; Process Control, chaired by Ruud Berkens and Robert Lowhon; and Emerging Areas, chaired by Larry McCloskey and Robert Thomas.

The papers presented at the conference were reviewed by the respective session chairs, and underwent minor editing by the conference director, before further editing and production by The American Ceramic Society.

Prof. Waltraud M. Kriven
Director, 64th Conference on Glass Problems
Department of Materials Science and Engineering,
The University of Illinois at Urbana-Champaign
March 2004

Acknowledgements

This conference would not have been possible without the continued involvement and voluntary efforts of the Program Advisory Committee, who shouldered the brunt of the technical organization. The Program Advisory Committee for the 64th conference on Glass Problems included:

Ruud G. C. Beerkens – TNO - TPD Glass Group, Eindhoven, The Netherlands

Daryl S. Clendenen – Vesuvius Monofrax, Inc., Falconer, NY

Thomas Dankert – Owens-Illinois, Toledo, OH

Charles H. Drummond III – The Ohio State University, OH

Marilyn deLong – Certainteed, Athens, GA

Robert Lawhon – PPG Industries, Pittsburg, PA

Larry McCloskey – Toledo Engineering Co., Inc., OH

Phil Ross – Glass Industry Consulting, Laguna Niguel, CA

Robert R. Thomas – Corning, Corning, NY

This committee identified and solicited forefront speakers on timely topics. The Program Advisory Committee has again ensured the high quality of papers presented at this conference, and their efforts are deeply appreciated. The contributing authors are gratefully acknowledged for sharing their knowledge with the glass community at large, in the form of their learned technical presentations and papers.

The conference was held at the Krannert Center for the Performing Arts. The official welcome to the conference was given by Dr. Bruce A. Vojak, Associate Dean for External Affairs of the College of Engineering. Prof. Ian M. Robertson, the Acting Head of the Department of Materials Science and Engineering made the opening remarks. The conference banquet was presided over by Prof. Charles H. Drummond III from The Ohio State University. Ms. Theresa Grentz, the UIUC women's basketball coach delivered a light-hearted, but pervasive after dinner speech.

Sincere and heartfelt appreciation goes to Mr. Jay Menacher, Assistant to the Head of the Department of Materials Science and Engineering, as well as to Ms. Mary Weaver, the Conference Secretariat at UIUC. Thanks to their tireless efforts, the logistics, facilities and services, as well as coordination and review of papers were smoothly and reliably managed.

Refractories

Surface Treatment of AZS Refractories Using High-Density Infrared Heating

T. N. Tiegs, F. C. Montgomery, D. C. Harper, and C. A. Blue
Oak Ridge National Laboratory, Oak Ridge, Tennessee

M. Velez, M. Karakus, and R. E. Moore
University of Missouri, Rolla, Missouri

Introduction

High-density infrared (HDI) technology is relatively new to the materials processing area and is gradually being exploited in coatings and surface modification. To date, it has been applied mainly to the treatment of metals.[1–3] However, recently it has been applied to the surface treatment of ceramic materials.[4] The HDI processing facility at ORNL uses a unique technology to produce extremely high-power densities of up to 3.5 kW/cm^2. Instead of using an electrically heated resistive element to produce radiant energy, a controlled and contained plasma is used. The advantages of the technology include the following:

1. Compared to laser technology, it can cover larger areas.
2. It consists of short-wavelength radiation (0.2–1.2 μm).
3. It has the ability to produce fast heating and cooling rates.
4. It is capable of attaining very high temperatures.
5. It has potential for continuous processing.

Prior studies have examined the surface modification of refractories using short-wavelength radiation.[5–9] For the most part, these studies have relied on laser melting of the surface. With that technology, the area actually being treated is quite small, with usual spot sizes of <6 mm in diameter. To surface treat a large area, the laser must be scanned across the part at speeds of 0.05–0.5 cm/s with typically a 20–50% overlay from the previous scan. To do a large area requires from several minutes up to several hours. In addition, the overlapping of the scans also affects the resulting microstructure and causes significant microcracking of the surface. Thus, this technique is of limited interest.

On the other hand, HDI is capable of much larger area coverage (up to 35 cm across) at reasonable scan speeds (~1 cm/s) making it a viable indus-

Table 1. Characteristics of AZS refractories

	FC AZS-33	CS AZS-20
Porosity (%)	3–9	17–20
Density (g/cm^3)	3.7	3.1
ZrO_2 (%)	33	20
SiO_2 (%)	13.5	10.7
Al_2O_3 (%)	52	69
Na_2O (%)	1.3	–
Others (%)	0.14	0.3

trial technique that is capable of continuously processing a large number of parts. Earlier efforts have shown that HDI processing can be applied to refractory materials to reduce surface porosity and reduce molten metal penetration.[10]

In the glass industry, alumina-zirconia-silica (AZS) refractories have been used for many years for glass contact applications. Previous work has shown that corrosion resistance is better with increasing zirconia content.[11] A study was initiated to demonstrate the fabrication of zirconia coatings on AZS refractories to improve the corrosion resistance.

Experimental Procedure

The HDI lamp consists of a 3.175-cm diameter quartz tube, which can be 11.5, 20, or 35 cm long. All of the tests on the refractories in the study used the 11.5 cm lamp. The quartz tube is cooled by a film of water on the inner diameter and a flow of argon through the tube. The plasma arc is generated between two tungsten electrodes inside the tube. (Further details on the operation of the system and the characteristics of the process can be found in Refs. 1–4.) The lamp is typically configured with a reflector to produce different areas of uniform irradiance and the distance between the lamp and the sample can be changed to vary the intensity of radiation impinging on the surface. The HDI lamp is mounted on a robotic manipulator arm so that it can be scanned across a surface. However, for the current study the lamp was operated in a stationary mode with the beam defocused to produce a uniform irradiance over the entire sample surface, and only the power to the lamp was varied to control the irradiation intensity at the sample surface.

The refractories treated in the study were commercial AZS materials: a fused cast AZS with 33% ZrO_2 (referred to as FC AZS-33) and a cast and sintered AZS with 20% ZrO_2 (referred to as CS AZS-20). The characteristics of each are presented in Table I. The samples measured approximately 1 × 1 × 2 in. (2.5 × 2.5 × 5 cm) and were cut from larger blocks. It is recognized that the sample surfaces may be slightly different than what would be found in an actual cast material, but for an initial study the samples were considered adequate. The coatings were made with slurries of ZrO_2 (3% Y_2O_3) powder and ethanol (10 vol% solids plus 1 wt% polyvinylpyrrolidone binder). The slurries were painted on by hand and dried. The weight of powder applied to each surface was estimated to produce a dense coating of zirconia ~200 μm thick.

Discussion of Results

HDI Treatment of Refractories

The surfaces of the refractories were exposed to the HDI treatment at various power levels. Several IR thermal treatments were done to determine an appropriate power level to use to bond the coating to the underlying refractory. Because of the high thermal expansion of the AZS refractories, the surface was preheated prior to the high-power bonding step. To determine the effects of the HDI treatment on the underlying refractory, several uncoated samples were treated using conditions similar to those for producing the coatings. In the case of the FC AZS-33, the surface exhibited a surface coating due to the exudation phenomenon typically observed when these types of refractories are heated. The sample also turned a buff color, which is also normal. On the other hand, the less dense CS AZS-20 sample showed very few morphological changes when HDI treated at conditions sufficient to produce melted coatings with the fused cast samples. The cast and sintered AZS required an additional IR exposure to produce a surface coating. This difference in behavior may be due to the higher SiO_2 and impurity contents in the dense FC AZS-33.

The appearances of the refractory samples at different stages are shown in Figs. 1 and 2 for the different refractories. Surface melting is readily evident for the HDI-treated samples. Weight changes observed during HDI treatments did not indicate that any severe decomposition took place. Cross sections of the HDI coatings are shown in Figs. 3 and 4. The HDI-treated surfaces on the FC AZS-33 revealed the melted surface coating was 1–2 mm thick. For the CS AZS-20, the HDI-affected zone was 0.5–1 mm thick.

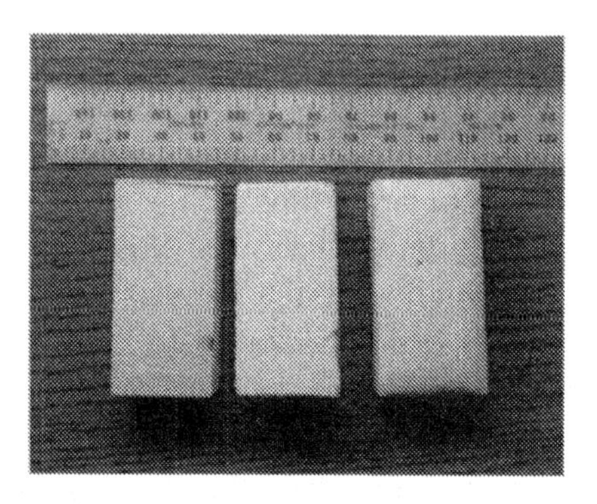

Figure 1. Visual appearance of FC AZS-33 before and after HDI treatment. Left to right: as-received sample; painted sample prior to HDI treatment; after HDI treatment. Surface melting was evident. HDI treatment consisted of a preheat from zero power to 220 W/cm^2 in 5 min, followed by a high-power treatment at 400 W/cm^2 for 4 s.

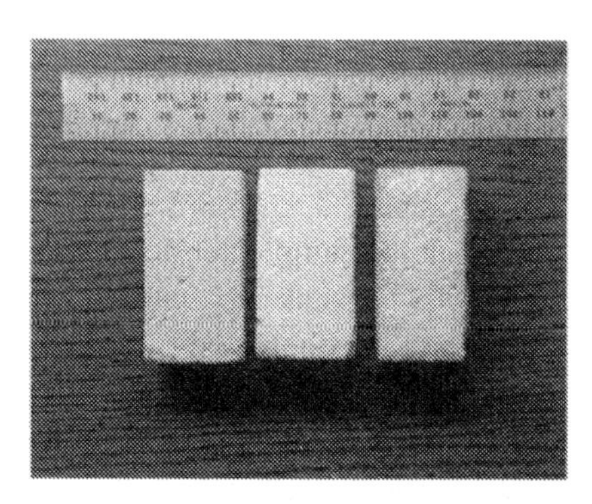

Figure 2. Visual appearance of CS AZS-20 before and after HDI treatment. Left to right: as-received sample; painted sample prior to HDI treatment; after HDI treatment. Surface melting was evident. HDI treatment consisted of a preheat from zero power to 220 W/cm^2 in 5 min, followed by a high-power treatment at 400 W/cm^2 for 4 s and an additional 375 W/cm^2 for 2 s.

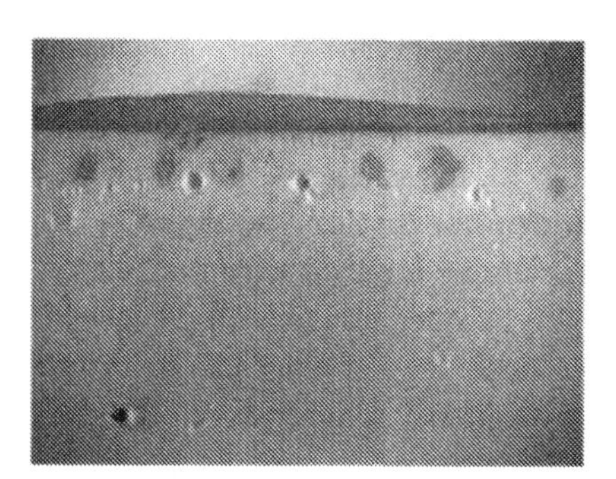

Figure 3. Cross section of coated FC AZS-33 after HDI treatment.

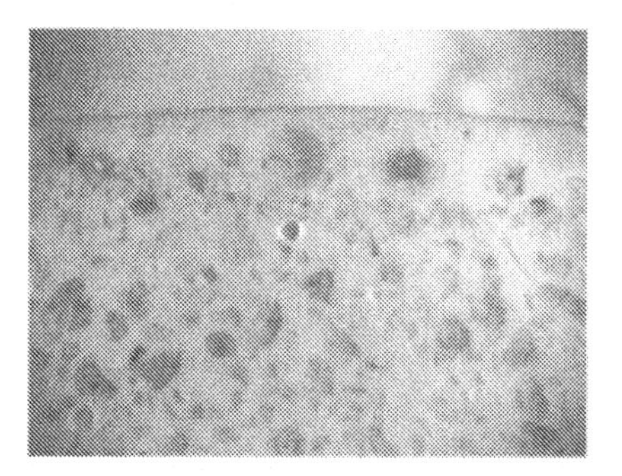

Figure 4. Cross section of coated CS AZS-20 after HDI treatment.

It should be noted that these thicknesses are considerably more than the 200-μm thick coatings that were estimated from the weight of zirconia powder that was applied to the surface prior to HDI processing. Evidently, significant mixing occurs between the applied zirconia powder coating and the underlying material. While the mixing could be viewed as detrimental, it probably helps to create a graded layer that would reduce the thermal

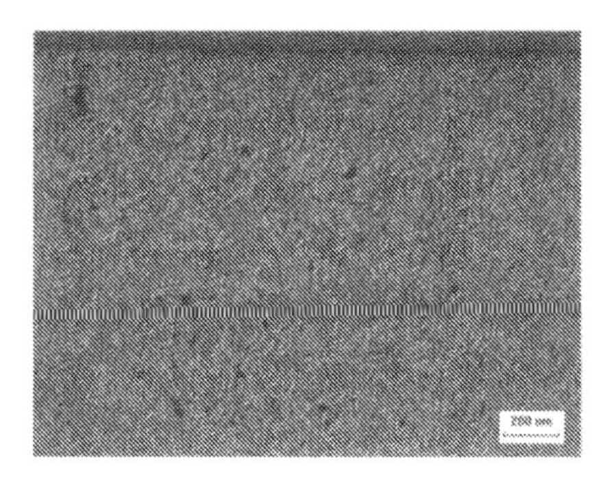

Figure 5. Microstructure of as-received FC AZS-33.

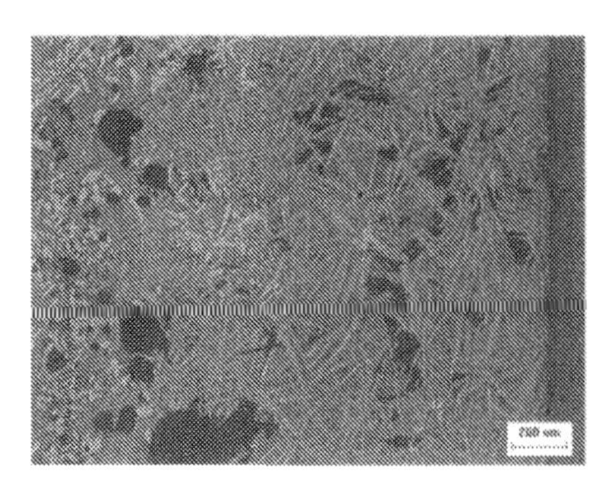

Figure 6. Microstructure of FC AZS-33 after HDI treatment.

Figure 7. Microstructure of as-received CS AZS-20.

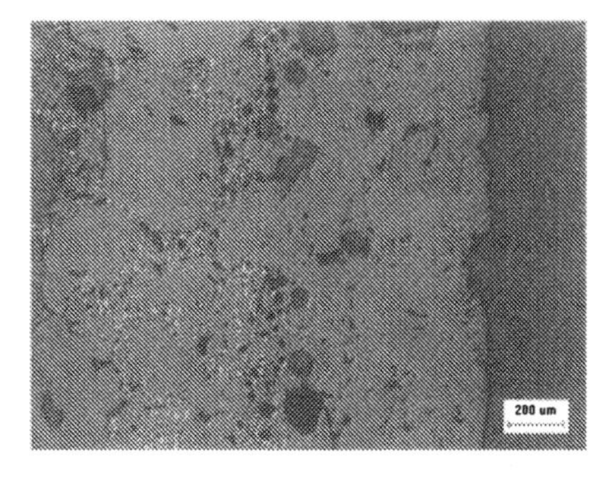

Figure 8. Microstructure of CS AZS-20 after HDI treatment.

expansion mismatch between the coating and refractory and should also improve the bonding.

The cross section also shows bubbles that are located generally near the interface between the melted region and the underlying material. A predominant source for the gas bubbles is believed to be internal porosity in the refractories that is trapped after the upper surface region becomes molten. Another source of the bubbles could be from trapped gases in the glass phase that have been reported for fused cast AZS refractories.

Significant microstructural changes occur during the HDI thermal treatments of the coated samples, as shown in Figs. 5–8. Normally, fused cast AZS microstructure consists of primary ZrO_2 (monoclinic) grains, Al_2O_3 grains with secondary ZrO_2, and a glassy silicate intergranular phase. The HDI-treated zirconia coatings in the present study were significantly differ-

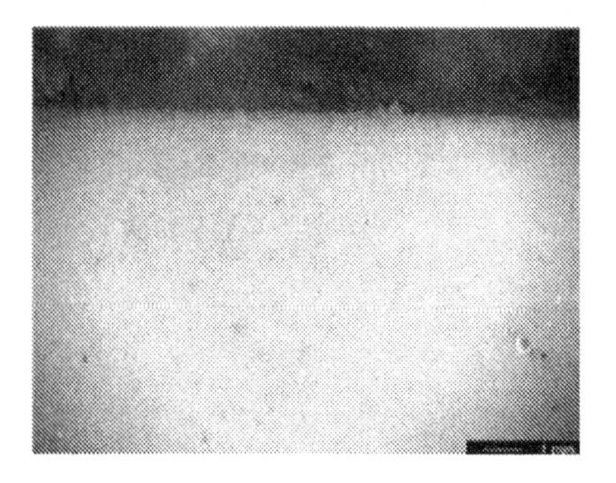

Figure 9. Cross section of as-received FC AZS-33 after corrosion testing for 15 h.

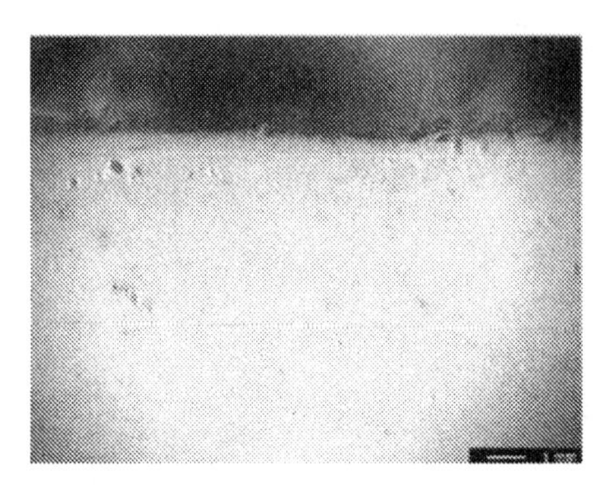

Figure 10. Cross section of HDI-coated FC AZS-33 after corrosion testing for 15 h.

ent. Evidently there is a melting of the zirconia coating as well as the top region of the underlying refractory. Mixing occurs during the liquid stage and high–aspect ratio ZrO_2 dendrites precipitate. X-ray diffraction identified tetragonal ZrO_2 in the HDI-treated coatings, as well as monoclinic. Porosity, which is probably due to shrinkage, is also observed. The microstructure of the as-received CS AZS-20 material shows large Al_2O_3 aggregates surrounded by a finer structure of ZrO_2 and aluminosilcate grains. HDI treatment of a coated sample showed melting and some mixing with the underlying refractory similar to that observed with the fused cast AZS. Dendrites of alumina and zirconia were observed in the HDI-affected surface region along with some residual Al_2O_3 aggregates from the underlying material. Porosity was also observed.

Corrosion Testing of HDI-Treated Refractories

Corrosion testing was done at 1400°C using a mixture of a typical soda-lime-silica window glass plus 10 wt% $NaCO_3$. It was thought that this mixture would be an aggressive test of the effectiveness of the coatings. Samples were exposed for either 15 or 80 h. Macroscopic views of the surfaces of the FC AZS-33 uncoated and coated samples after 15 h showed some degradation of the refractories (Figs. 9 and 10). Closer inspection revealed the formation of a boundary layer consisting of undissolved ZrO_2 grains, which is normal behavior for this type of refractory (Figs. 11 and 12). For the uncoated FC AZS-33, the boundary layer was ~300 μm. For the coated FC AZS-33, the boundary layer was not readily apparent. However, disso-

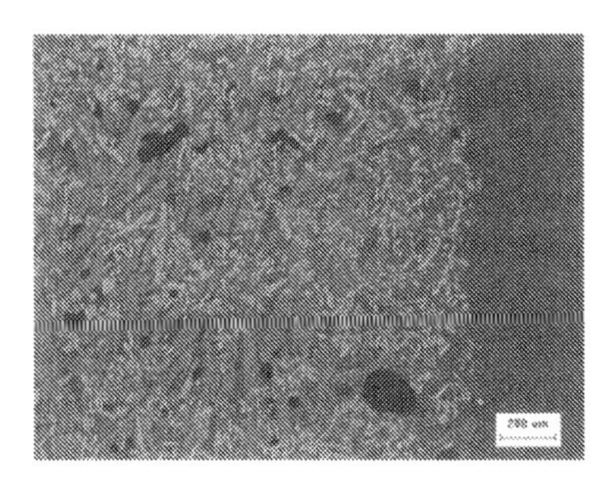

Figure 11. Microstructure of as-received FC AZS-33 after corrosion testing for 15 h.

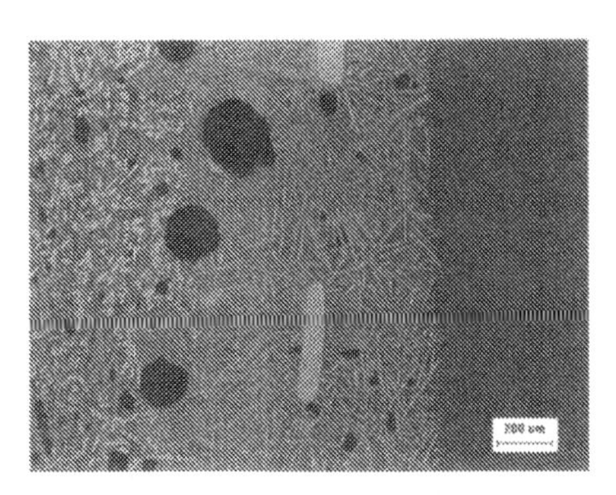

Figure 12. Microstructure of HDI-coated FC AZS-33 after corrosion testing for 15 h.

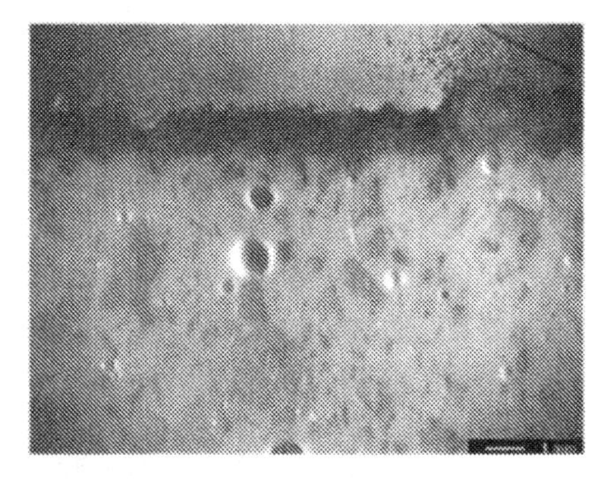

Figure 13. Cross section of as-received CS AZS-20 after corrosion testing for 15 h.

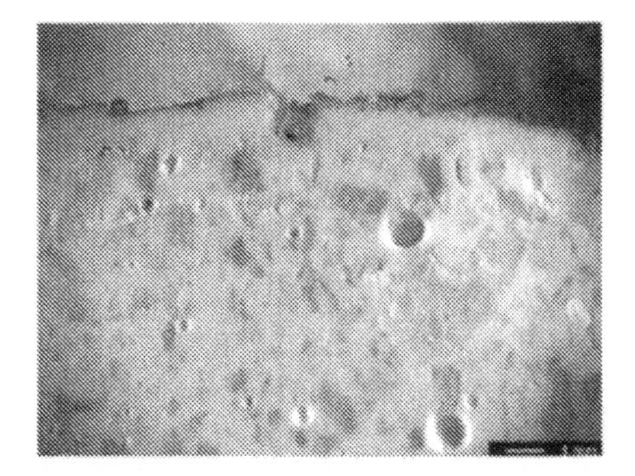

Figure 14. Cross section of HDI-coated CS AZS-20 after corrosion testing for 15 h.

lution of the intergranular phase between the high–aspect ratio ZrO_2 dendrites appeared to have occurred to a depth of ~300 μm below the surface. The CS AZS-20 also showed degradation during the 15-h glass mixture exposure (Figs. 13 and 14). The uncoated sample exhibited significant dissolution and release of undissolved grains as shown in Fig. 15. The coated CS AZS-20 showed some degradation; however, a boundary layer of undissolved ZrO^2 was still observed on the surface (Fig. 16).

As expected, the 80-h exposure tests revealed significantly more degradation of the refractories (Figs. 17 and 18). On the uncoated FC AZS-33 materials, the boundary layer had a thickness of ~900 μm and a considerable amount of undissolved ZrO_2 grains within the glass melt due to materi-

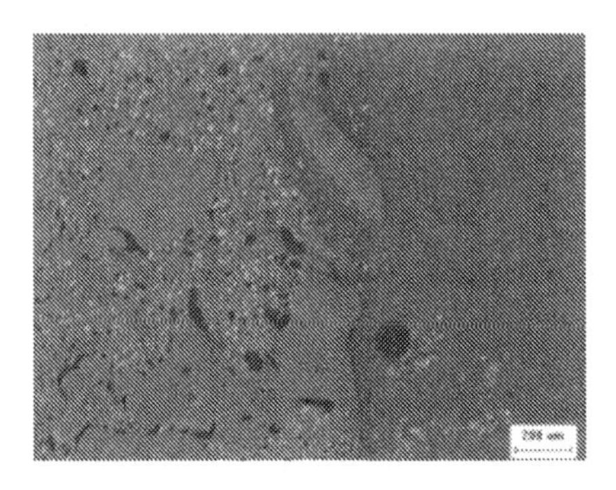

Figure 15. Microstructure of as-received CS AZS-20 after corrosion testing for 15 h.

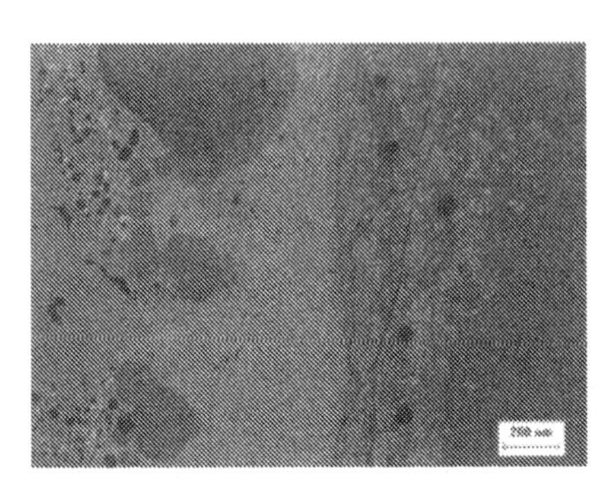

Figure 16. Microstructure of HDI-coated CS AZS-20 after corrosion testing for 15 h.

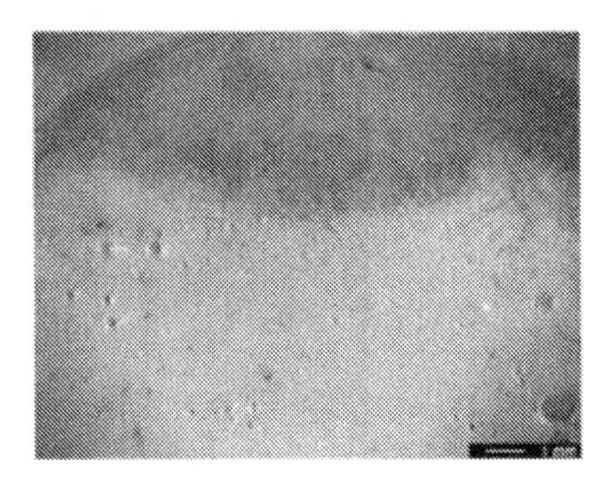

Figure 17. Cross section of as-received FC AZS-33 after corrosion testing for 80 h.

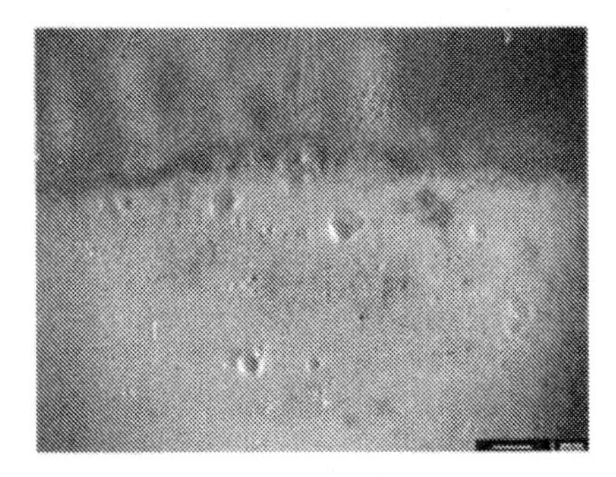

Figure 18. Cross section of HDI-coated FC AZS-33 after corrosion testing for 80 h.

al loss from surface (Fig. 19). The coated material showed a larger boundary layer due to retention of fine ZrO_2 from the HDI coating, a significant dissolution of the alumina grains and glassy grain boundary phase in the underlying refractory (Fig. 20).

All CS AZS-20 materials showed severe degradation after the 80-h glass mixture exposure (Figs. 21 and 22). In fact, the HDI-coated sample had the coating on one surface, and those areas not coated were nearly disintegrated during the test. Microscopic examination of the uncoated CS AZS-20 reveals the severity of the degradation with both fine ZrO_2 and larger Al_2O_3 grains being dislodged from the surface (Fig. 23). The HDI-coated surface of the CS AZS-20 showed the retention of the fine ZrO_2 in the coating forming a significant boundary layer (Fig. 24).

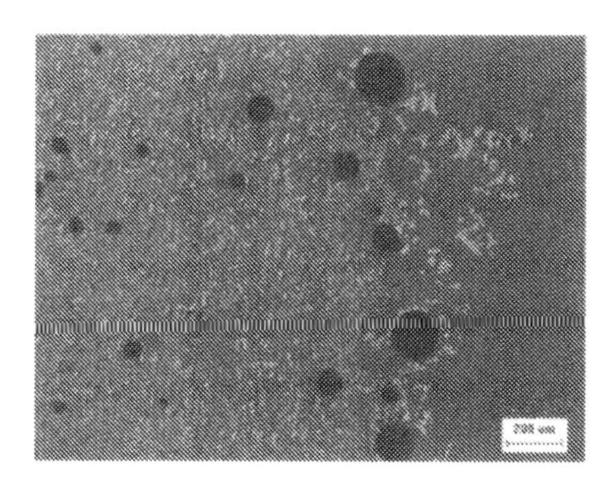

Figure 19. Microstructure of as-received FC AZS-33 after corrosion testing for 80 h.

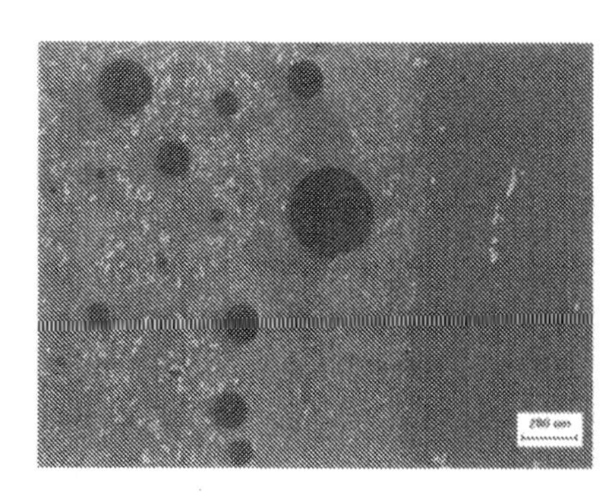

Figure 20. Microstructure of HDI-coated FC AZS-33 after corrosion testing for 80 h.

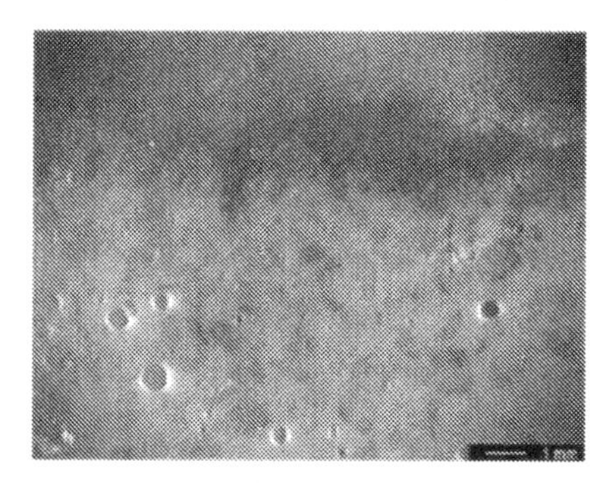

Figure 21. Cross section of as-received CS AZS-20 after corrosion testing for 80 h.

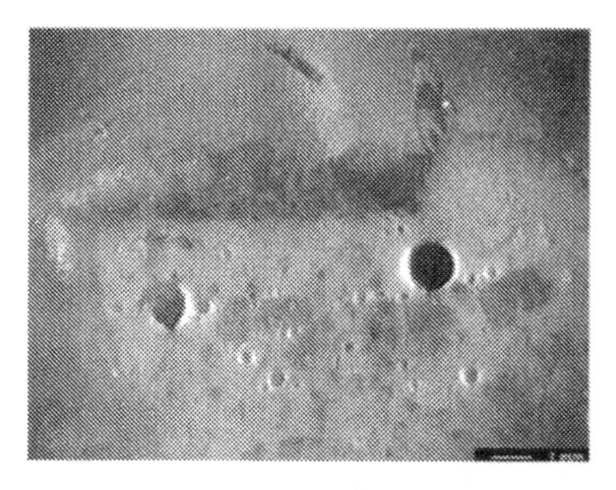

Figure 22. Cross section of HDI-coated CS AZS-20 after corrosion testing for 80 h.

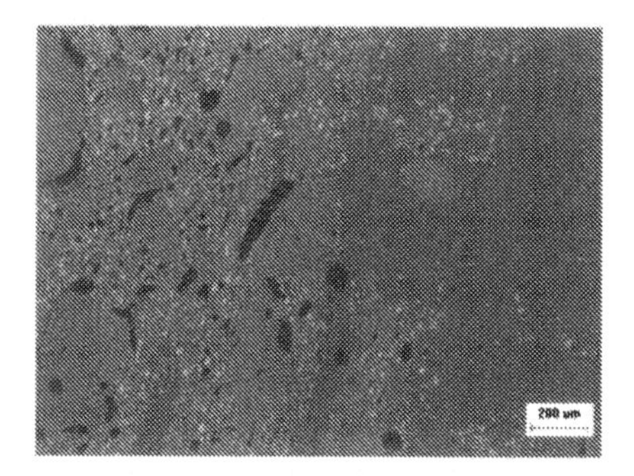

Figure 23. Microstructure of as-received CS AZS-20 after corrosion testing for 80 h.

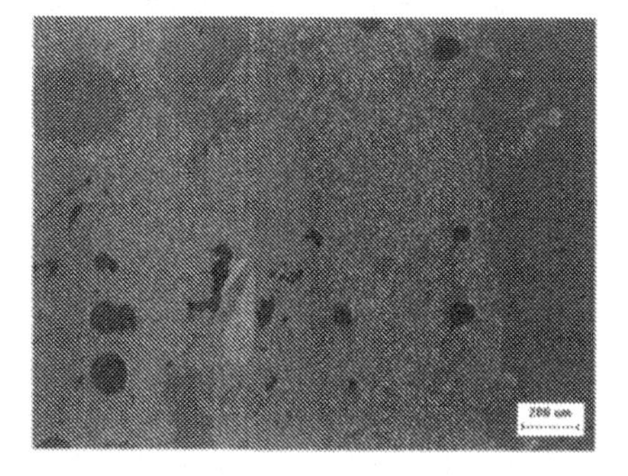

Figure 24. Microstructure of HDI-coated CS AZS-20 after corrosion testing for 80 h.

Conclusions

It was demonstrated that zirconia-rich coatings can be formed on the surfaces of AZS refractories using HDI thermal treatments. During high-temperature processing, the surfaces are melted and mixing occurs between the applied zirconia coating and the underlying refractory. The coatings are characterized by fine dendritic ZrO_2 grains and some porosity. Corrosion testing in a container glass–$NaCO_3$ mixture at 1400°C showed that the coatings help form boundary layers on the surfaces.

Acknowledgments

This research was sponsored by the U.S. Department of Energy, Office of Energy Efficiency and Renewable Energy, Industrial Technologies Program, Industrial Materials for the Future under contract DE-AC05-00OR22725 with UT-Battelle, LLC.

References

1. C. A. Blue, "High-Intensity Lamp Opening New Surface-Treating Vistas," *Industrial Heating,* March 2002, pp. 79–82.
2. C. A. Blue et al., "High-Density-Infrared Transient Liquid Coatings," *J. Metals-e,* **52** [1] (2000).
3. O. C. Meyer-Kobbe, "Surface Treatment with a High-Intensity Arc Lamp," *Adv. Mater. Proc.,* **9** (1990).
4. T. N. Tiegs, J. O. Kiggans, F. C. Montgomery, C. A. Blue, and M. Velez, "High-Density Infrared Surface Treatment of Ceramics," *Am. Ceram. Soc. Bull.,* **82** [2] 49–53 (2003).
5. A. Petitbon et al., "Laser Surface Treatment of Ceramic Coatings," *Mater. Sci. Eng.,* **A121,** 545–548 (1989).
6. S. Bourban et al., "Solidification Microstructure of Laser Remelted Al_2O_3-ZrO_2 Eutectic," *Acta Mater.,* **45** [12] 5069–5075 (1997).
7. L. Bradley et al., "Characteristics of the Microstructures of Alumina-Based Refractory Materials Treated with CO_2 and Diode Lasers," *Appl. Surf. Sci.,* **138–139**, 233–239 (1999).
8. L. Bradley et al., "Surface Modification of Alumina-Based Refractory Materials Using a Xenon Arc Lamp," *Appl. Surf. Sci.,* **154–155**, 675–681 (2000).
9. L. Bradley et al., "Flame-Assisted Laser Surface Treatment of Refractory Materials for Crack-Free Densification," *Mater. Sci. Eng.,* **A278,** 204–212 (2000).
10. T. N. Tiegs, J. O. Kiggans, F. C. Montgomery, D. C. Harper, and C. A. Blue, "Surface Modification of Ceramics by High-Density Infrared Heating," *Ceram. Eng. Sci. Proc.,* **24** [4] (2003).
11. M. Dunkl et al., "Formation of Boundary Layers on Different Refractories in Glass Melts," *Ceram. Eng. Sci. Proc.,* **24** [1] 211–223 (2003).

Investigation of Defects in High-Quality Glasses

K. R. Selkregg and A. Gupta
Vesuvius Monofrax Inc.

Introduction

Glass defects occur in many forms — for example, bubbles, stones, knots, and cords. Glass defects originate from many sources, such as poor melting and/or refining of glass, refractory quality, and degradation. The focus of this paper is on viscous knots and cords originating from AZS refractories with a special emphasis on the chemistry and frequency of these defects as a function of furnace age and refractory reuse. Factors such as furnace throughput and temperature profile/stability on defects dissolution are not considered in this study.

It is evident that defect identification through chemistry and microscopy has been described extensively in literature (see bibliography). Even though the role of AZS exudation and corrosion on glass defects has been described, some confusion still persists on the long-term source of defects. The published literature often cites AZS refractory exudation as the leading cause of knot and cord defects in glass. Monofrax presented a paper on the origin of knot defects in TV glass at the Glass Problems Conference in 2001. In this paper, it was shown that AZS refractory exudation is only a short-term source of glass defects, whereas AZS refractory corrosion, especially in the superstructure, is a more potent and long-term source of defects.

There is very little information in the literature on defect chemistry and frequency as a function of furnace age and AZS reuse.

Results from variable-term laboratory and field experiments will be reviewed. In addition, post-campaign AZS refractory blocks from three types of glass melting furnaces (soda-lime container, soda-lime tubing, and lead silicate TV funnel furnaces) have been analyzed in order to shed some light on the chemistry and frequency of defects as a function of furnace age.

Figure 1

Before we proceed with the results of the various AZS refractory analyses mentioned above, it is important to briefly review the AZS refractory

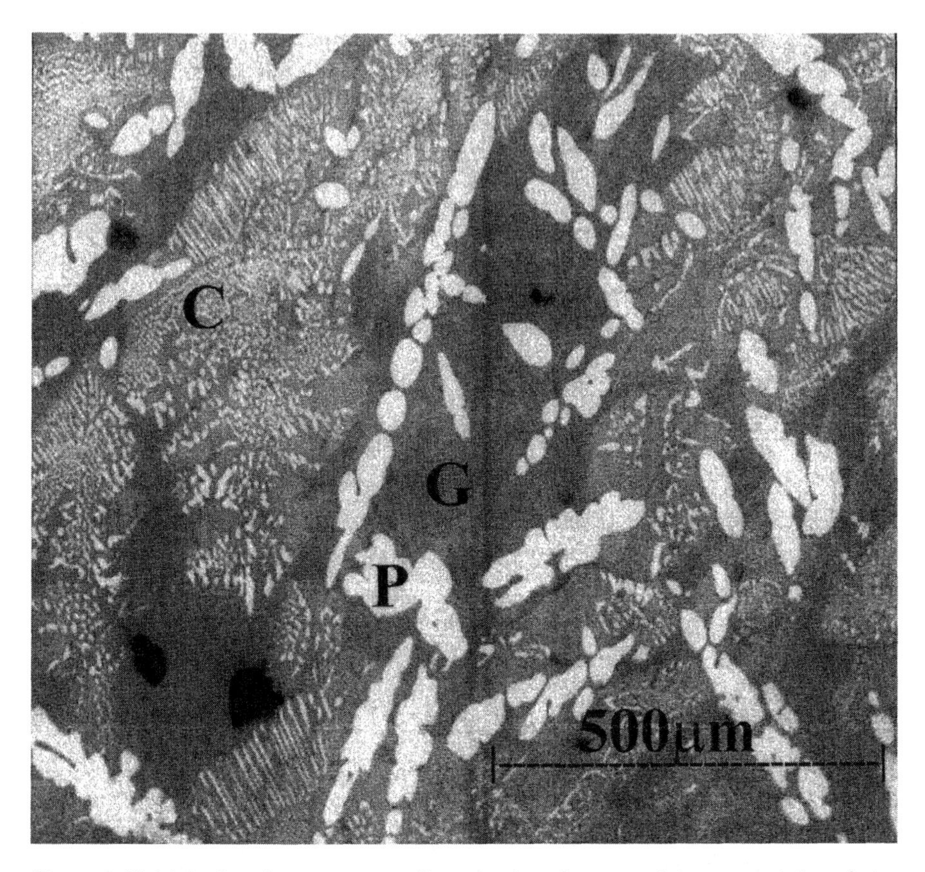

Figure 1. Multiphasic refractory: crystalline alumina plate containing coprecipitated zirconia crystals (C), independent dendritic zirconia (P), and glassy matrix phase (G). The diffusive path of corrosive species in the furnace environment is through the matrix (glassy) phase. Analysis of matrix phase chemistry as a function of depth is a sensitive indicator of the refractory's chemical alteration.

microstructure and the mechanism of corrosion. The AZS refractory consists of three major phases as shown in Fig. 1. The glass phase, which accounts for 30–35% of the total volume of the AZS refractory, provides the primary path for in-diffusion of corrosive species from both the molten glass bath as well as the gaseous environment in the furnace above the glass bath. The depth of penetration of the corrosive species into the glassy matrix phase of AZS refractory, with attendant changes in refractory microstructure, is often the best indicator of the overall refractory corrosion.

Figure 2

Figure 2 shows the results of a short-term glass contact corrosion test performed in the Monofrax laboratory using a TV panel glass.

Figure 2(b) shows the concentration of the major components of the glassy matrix phase as a function of depth into the refractory. These data were collected with a scanning electron microscope in conjunction with an energy dispersive spectrometer, at known depths into the AZS refractory.

Figure 2(a) shows the microstructure of the AZS refractory at the glass/refractory interface following the corrosion test. This micrograph shows a near absence of crystalline alumina to

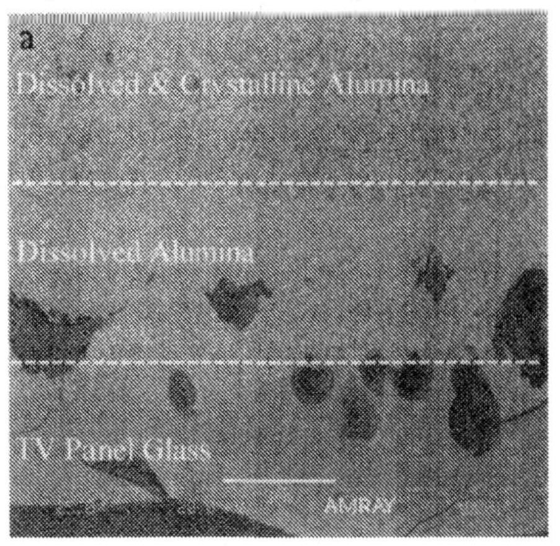

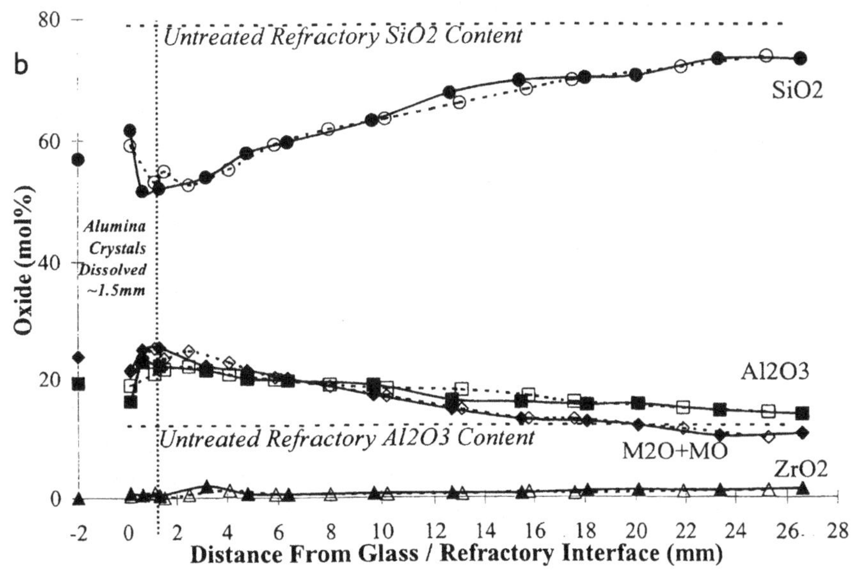

Figure 2. Short-term corrosion of AZS. (A) SEM/BSE image; (B) SEM/EDS. Melt contact corrosion was at 1560°C and 150 h. Evidence of alkali dissolution of crystalline alumina is observed up to ~25 mm.

a depth of ~2 mm from the interface. This region of the AZS refractory is typically referred to as the passivation or boundary layer and consists of an Al_2O_3-rich glassy phase and ZrO_2 crystals. Although this passivation layer serves to impede the corrosion rate of the AZS refractory, any mechanical force, such as convection currents at the glass melt–refractory–vapor phase (also known as metal line) or AZS exudation, can disrupt this protective layer and promote formation of glass defects.

Figure 2(b) shows the familiar changes in chemistry of the AZS glassy phase chemistry. There is an increase in the concentration of the alkaline and alkaline-earth species accompanied with an increase in the Al_2O_3 concentration. The SiO_2 concentration in the glassy phase actually decreases toward the glass/refractory interface even as the total volume of the glassy phase increases. These changes in the glassy phase components are most noticeable to a depth of ~25 mm. The symbols plotted directly on the vertical axis show the chemistry of a knot defect obtained from a TV panel glass furnace. By comparing the chemistry of the defect with the chemistry of the AZS refractory at the glass/refractory interface, the defect source can be established.

Figure 3

Figure 3 shows the results from superstructure corrosion of 34% ZrO_2 AZS refractory in a lead silicate crystal glass furnace following two separate test durations. Here again, the corrosion of the AZS refractory is expressed in terms of the depth to which the glassy matrix phase chemistry has been altered. While the nature of the chemical change in the glassy phase chemistry is similar following the two corrosion tests, the depth of alteration is clearly greater following the longer term test.

Figure 4

The next several figures will show the results from evaluation of post-campaign AZS refractory blocks from industrial glass melting furnaces. Figure 4 shows photos of an AZS superstructure and a glass contact refractory block taken from a soda-lime container furnace following a 10-year campaign. While the glass contact block, right in the photo, shows rounding of the block edges due to corrosion and a shiny surface due to glass adhering to the refractory, the superstructure refractory appears to be dry on the surface, has relatively sharp edges, and shows a whitish crust on the entire

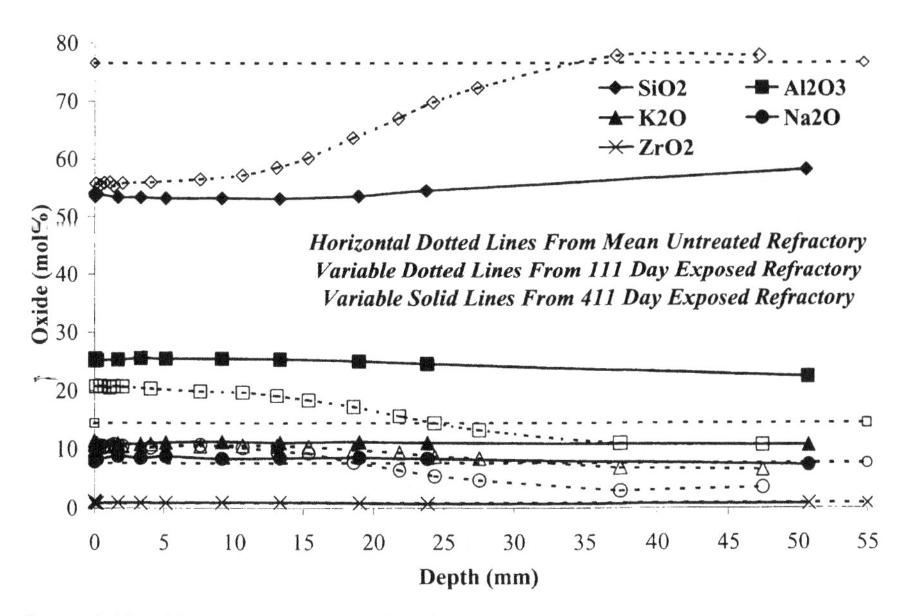

Figure 3. Variable-term corrosion of AZS. Superstructure corrosion in lead silicate at 111 and 411 days. Glass phase composition in exposed Monofrax CS3 AZS. Corrosion depth varies from 35 mm to at least 50 mm.

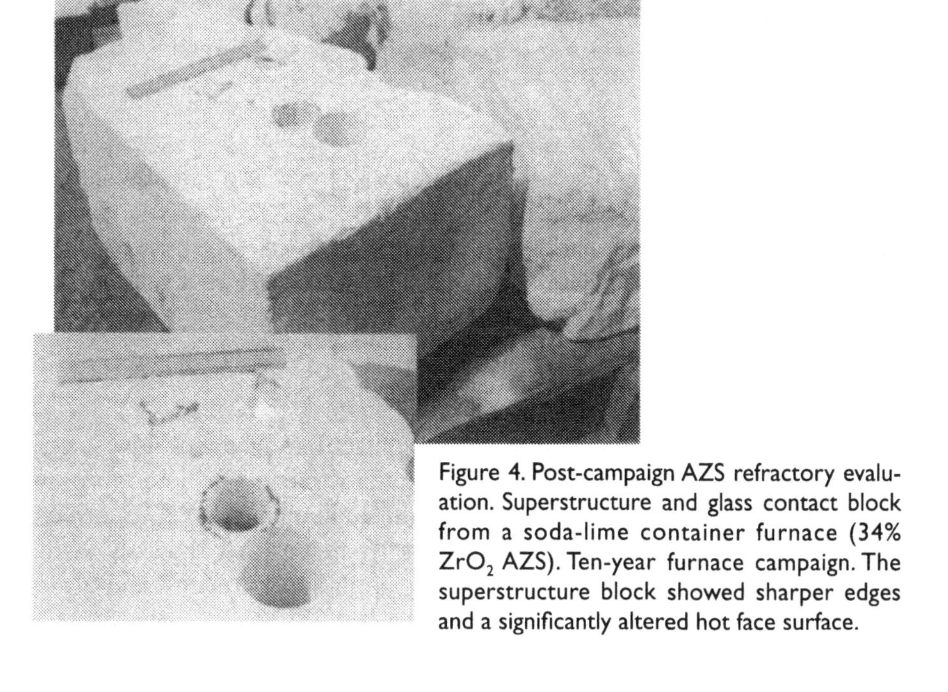

Figure 4. Post-campaign AZS refractory evaluation. Superstructure and glass contact block from a soda-lime container furnace (34% ZrO_2 AZS). Ten-year furnace campaign. The superstructure block showed sharper edges and a significantly altered hot face surface.

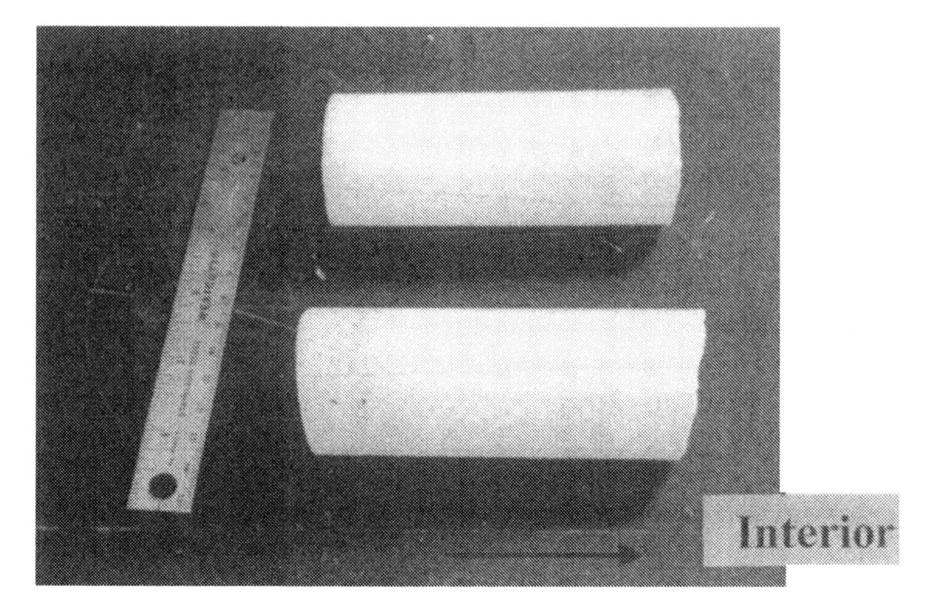

Figure 5. Post-campaign AZS refractory evaluation. Core samples analyzed to document chemical, physical, and microstructural changes as a function of depth.

exposed hot face. The glass contact block appears to have come from below the metal line. The holes in the blocks are from drilling core samples for characterization. The results are shown in the following figures.

Figure 5

Polished sections for microscopy were prepared from core samples shown in Fig. 5. The chemistry and microstructure were analyzed as a function of depth using SEM/EDS techniques. In addition, physical properties such as bulk density and apparent porosity were also measured as a function of depth.

Figure 6

SEM/BSE photomicrographs of the glass contact interface and superstructure hot face are shown in Fig. 6(a). The microstructure of both types of samples shows a near absence of the crystalline alumina phase. This is very similar to the observations made in the earlier figures that show short- to medium-term laboratory corrosion results. Both superstructure and glass

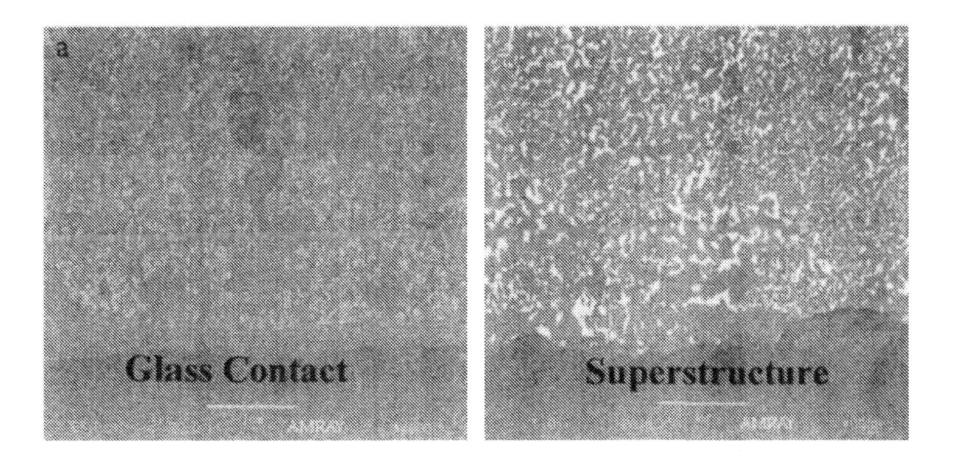

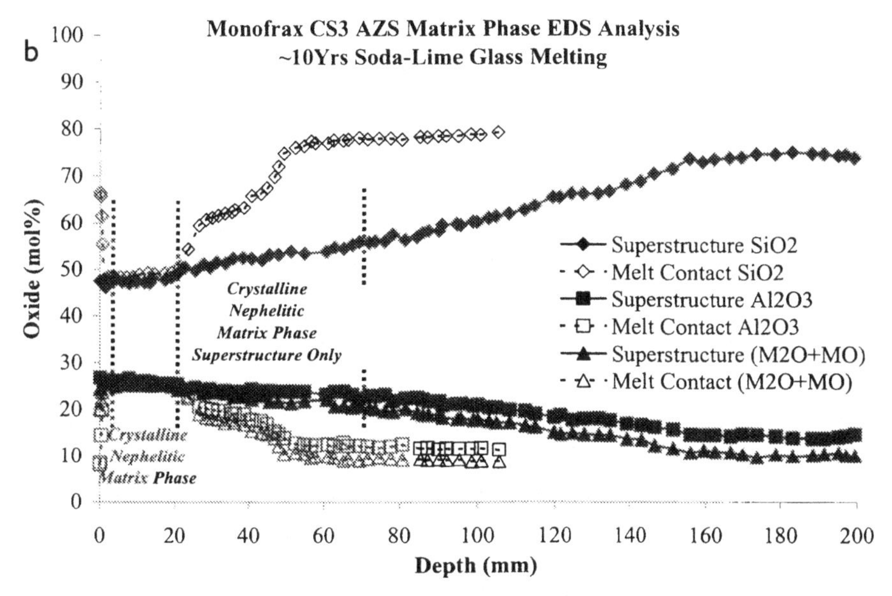

Figure 6. Post-campaign glass contact and superstructure AZS corrosion; soda-lime container glass (12.9 mol% Na_2O, 2.3 mol% MgO, 0.9 mol% Al_2O_3, 72.3 mol% SiO_2, 0.2 mol% K_2O, 11.6 mol% CaO), 10-year campaign. The corrosion depth was ~150 mm in the superstructure block, compared to ~50 mm at glass contact.

contact AZS samples were found to contain a nephelitic zone where the zirconia is no longer in solution in the glassy phase; instead it exists as discrete zirconia crystals. This nephelitic zone extends to greater depth in the superstructure than in the glass contact.

Figure 6(b) shows the matrix phase chemistry as a function of depth for both the glass contact and superstructure AZS refractory samples. Both samples exhibit similar changes in chemistry, that is, an increase in the concentration of the alkali/alkaline earth species, an increase in the alumina concentration, and a decrease in the concentration of silica. The superstructure refractory sample, however, shows a greater depth of chemical change than the glass contact sample. An example of this is the depth of the nephelitic zone, mentioned above, which was found to be greater in the superstructure sample than in the glass contact sample.

The results shown in this figure suggest that the glass contact refractory corrosion passivates with time, whereas the superstructure corrosion continues over the life of the furnace campaign.

Figure 7

Figure 7 shows a photo of an AZS glass contact sidewall block obtained from a soda-lime glass tubing furnace following a 5-year campaign. The analysis of this block is our second industrial case study. The photo on the bottom shows a viscous knot glass defect also obtained from this tubing furnace. As well as analyzing the chemical, physical, and microstructural changes in the refractory block, the glass defect was also analyzed to determine if its source was the glass contact AZS refractory.

Figure 8

The SEM/BSE photomicrograph shown in Fig. 8 displays the microstructure of the AZS block at the glass/refractory interface. The changes in the AZS microstructure are similar to those described in earlier figures. The crystalline alumina phase is essentially absent in the micrograph shown.

The chart shows the chemical analysis of the matrix phase (glassy in the as-is AZS refractory) as a function of depth in the AZS block. The chemistry of this tubing glass is similar to that of the container glass discussed earlier. The chemistry profile of the AZS block was found to be similar to that seen in the container glass tank. The alkali/alkaline earth and the alumina contents increase, whereas the silica content decreases in the matrix toward the glass/refractory interface. This resulted in formation of a nephelitic matrix phase.

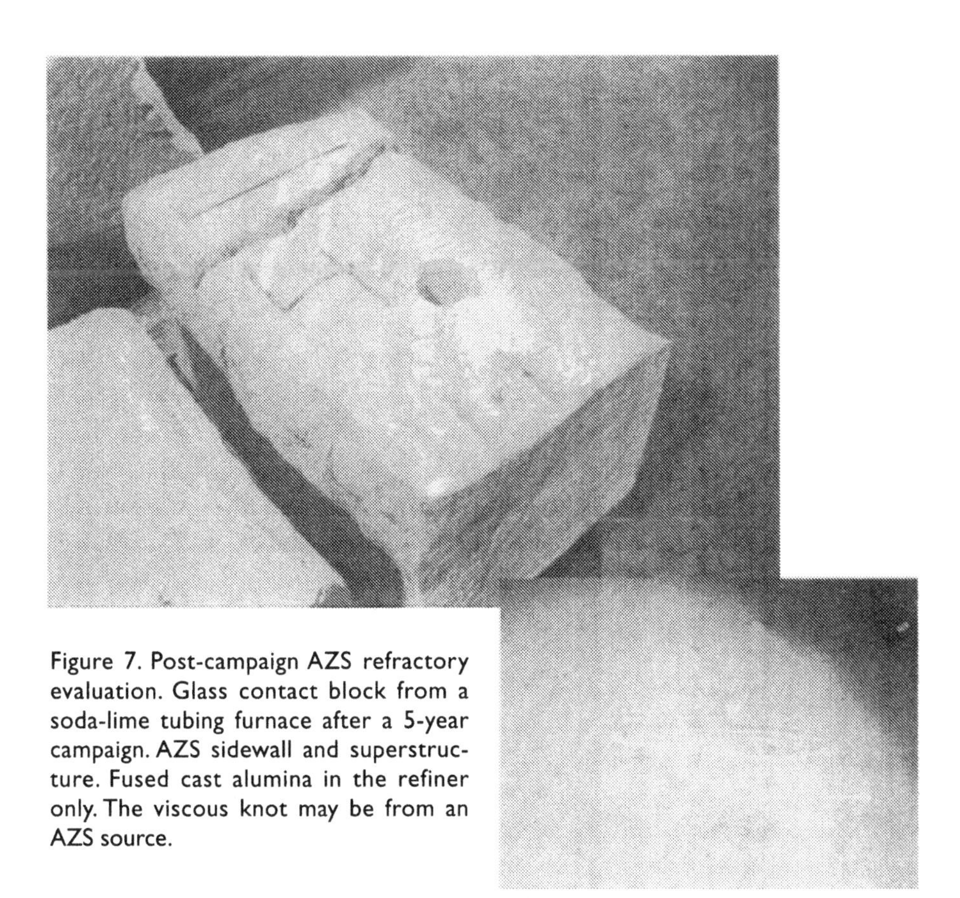

Figure 7. Post-campaign AZS refractory evaluation. Glass contact block from a soda-lime tubing furnace after a 5-year campaign. AZS sidewall and superstructure. Fused cast alumina in the refiner only. The viscous knot may be from an AZS source.

The chart also shows the chemistry of the viscous knot defect, represented by open symbols at an arbitrary position of –5 mm from the glass/refractory interface. The lack of ZrO_2 in the cord chemistry strongly suggests a non-AZS cord source.

Figure 9

This figure compares the depth of corrosion in the glass contact AZS blocks taken from the container and the tubing furnaces. Although the corrosion depth is greater in the container furnace, this could be due to the differences in the campaign durations (10 years for the container vs. 5 years for the tubing), location in the furnace, and temperature.

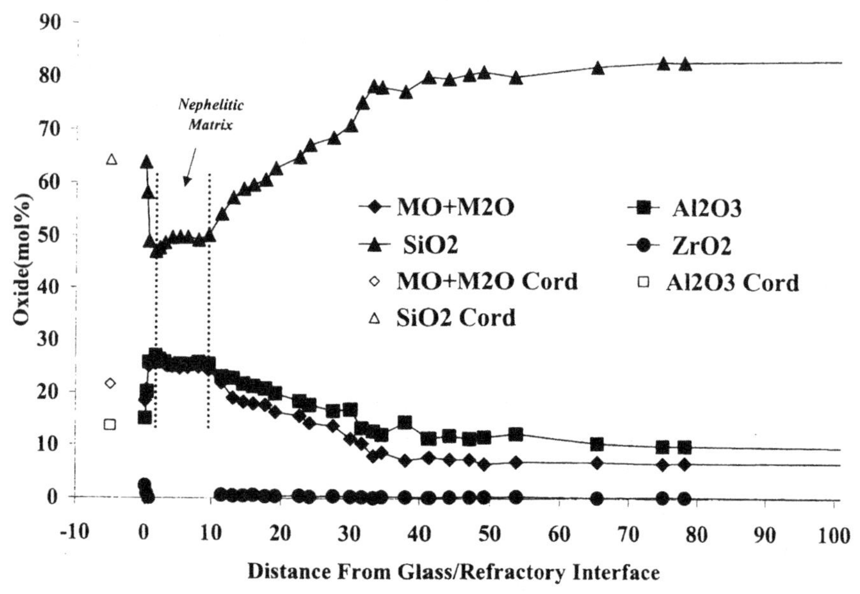

Figure 8. Post-campaign glass contact and superstructure AZS corrosion; soda-lime tubing (15.5 mol% Na_2O, 4.7 mol% MgO, 0.9 mol% Al_2O_3, 73.5 mol% SiO_2, 0.25 mol% K_2O, 5.2 mol% CaO), 5-year campaign. The corrosion depth was ~35 mm. Open symbols represent cord chemistry; the absence of ZrO_2 in the cord is indicitive of a non-AZS source.

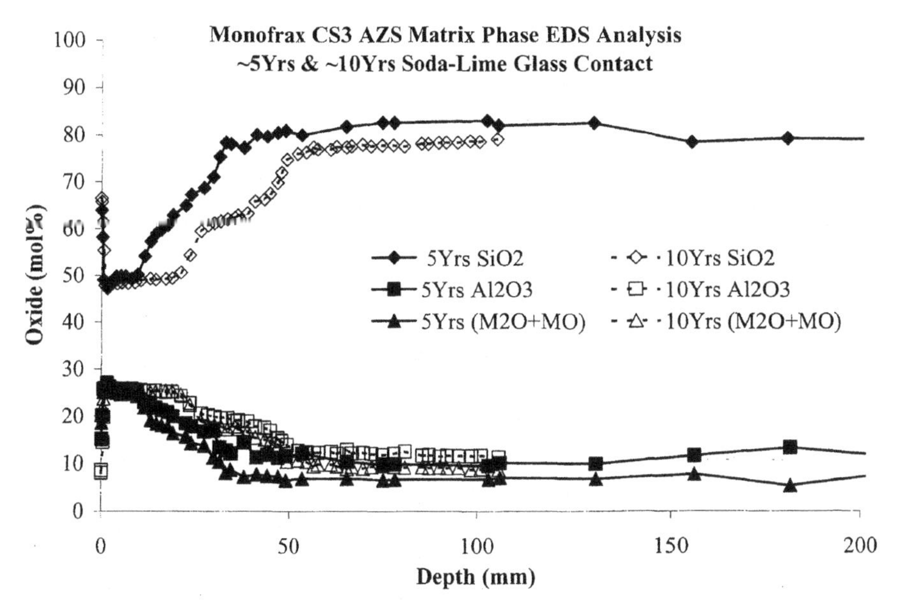

Figure 9. Post-campaign glass contact AZS corrosion: soda-lime furnaces, 5- and 10-year comparison. Monofrax CS3 AZS matrix phase EDS analysis. Corrosion depth was greater in the 10-year campaign AZS block. However, location in the furnace and temperature could also explain the difference.

Figure 10

The third post-campaign AZS refractory sample was obtained from a TV funnel furnace after ~4 years of service. A core sample, B1 as shown in Fig. 10, was drilled from the center of the charge end superstructure wall. Extensive rundown can be seen on the wall. This is most likely due to the corrosion of the AZS blocks, accelerated by the inevitable presence of batch dust in the charge end area. A cross section of the core sample shows a tear running from the hot face into the block interior, and a change in the color of the refractory.

Figure 11

Figure 11 displays the concentration of all alkali and alkaline earth species in the B1 core sample as a function of depth. The chemical analysis was performed using SEM/EDS in the line scan mode at a low magnification.

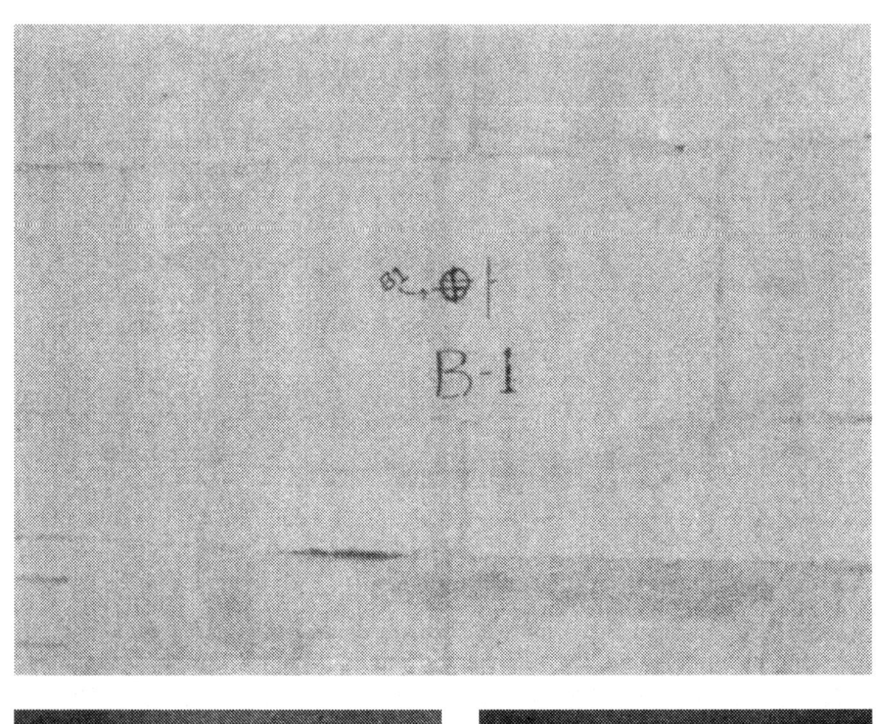

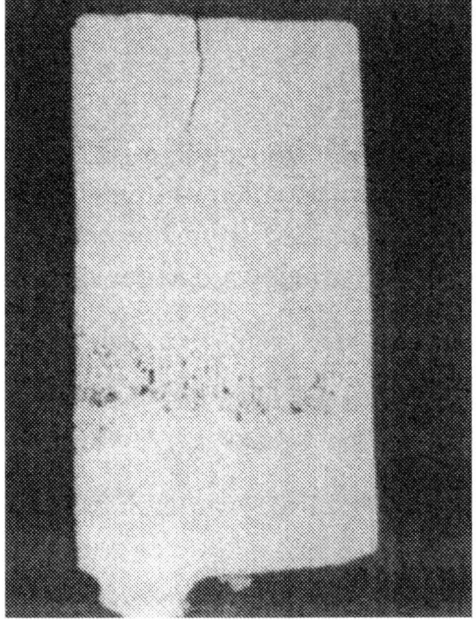

Figure 10. Post-campaign AZS refractory evaluation. Superstructure core sample from a TV funnel furnace charge end wall, ~4-year furnace campaign. Note the extensive rundown and change in internal color.

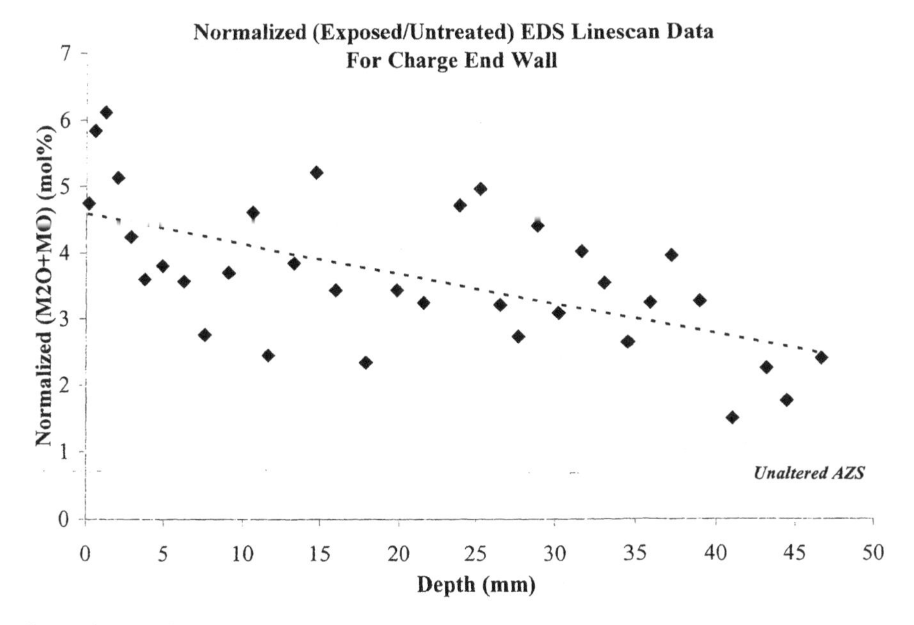

Figure 11. AZS superstructure corrosion in a TV funnel furnace after 4 years of service. There was significant penetration of alkali and alkaline earth species into AZS refractory up to at least 50 mm, creating conditions for crystallization of matrix phase.

This method of analysis allows bulk chemical analysis at a known depth into the sample. The chart compares the alkali and alkaline earth concentration of the exposed AZS sample to the concentration of the same species in as-is AZS refractory. There is a significantly higher concentration of the analyzed species up to at least 50 mm.

Figure 12

A series of reflected light photomicrographs (Fig. 12) were collected from the polished samples used to generate the results shown Fig. 11. A complex array of phases has formed because of the chemical changes in the AZS block. The in-diffusion of alkaline and alkaline earth species (potassium, sodium, and magnesium) with the concomitant dissolution of crystalline alumina into the matrix has resulted in creation of new crystalline phases such as nepheline, kalsilite, leucite, and β-alumina.

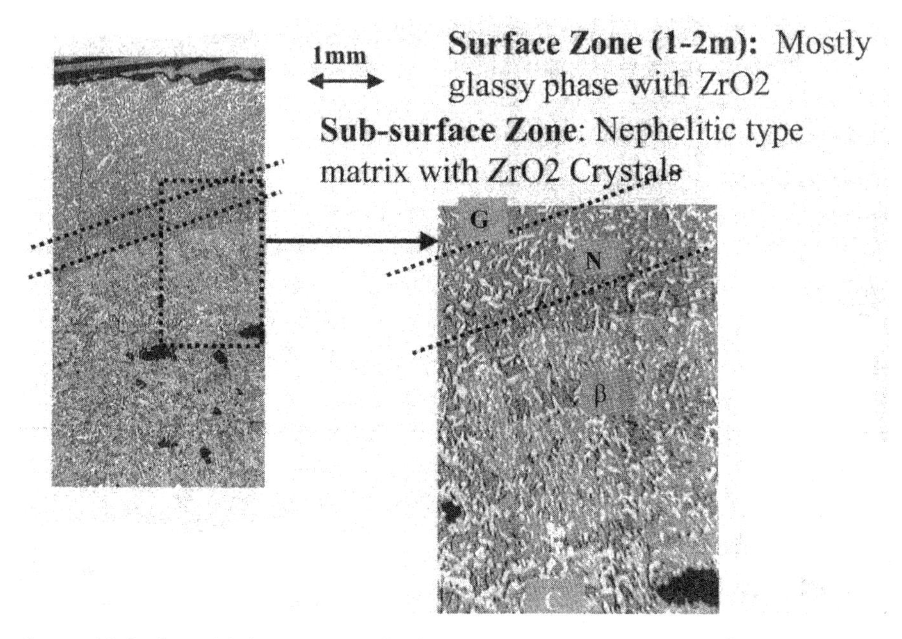

Figure 12. Reflected light images of TV funnel charge end AZS. (Mg,K) β-alumina plates replaced corundum during service.

Table I. Summary of AZS corrosion depth results

	Short-term (< 1 month)	Medium-term (1 month to 1 year)	Post-campaign (years)
Glass contact	~25 mm TVP		50 mm SLC 35 mm SLT
Superstructure	~15 mm TVP	40–50+ mm lead crystal	150 mm SLC >50 mm TVF

Table I

Table I summarizes the depth of corrosion measured in the all of the AZS samples discussed in this paper. While the numbers shown in the table are not meant to be used for calculating the rate of corrosion of glass contact and superstructure AZS refractories, the trend observed here supports the conclusion that superstructure AZS undergoes greater degree of chemical change than the glass contact AZS below the metal line. All other things being equal, these data therefore suggest that superstructure corrosion can

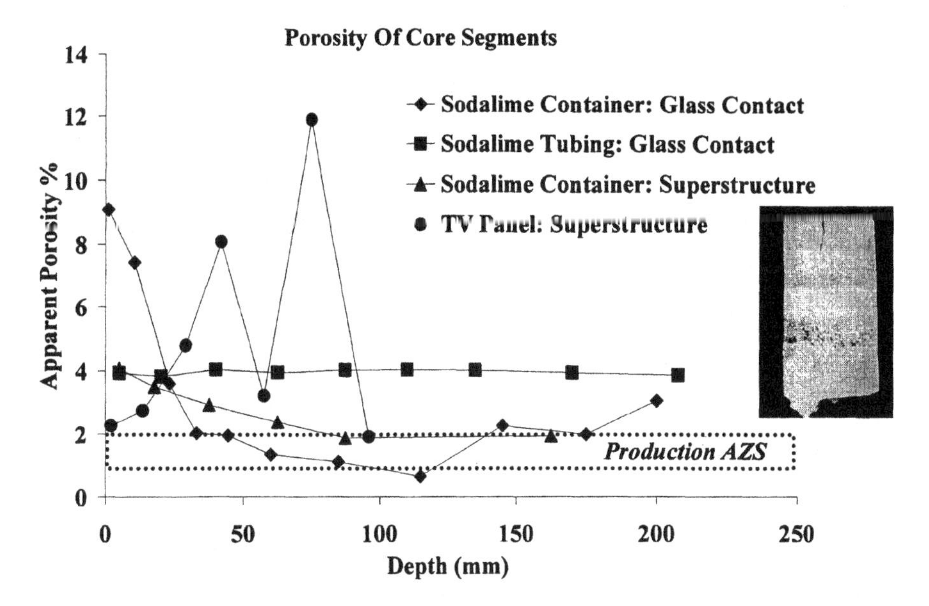

Figure 13. AZS refractory contains 1–2% apparent porosity. Post-campaign data show higher porosity. This could be due to phase changes and/or matrix (liquid) phase rundown.

continue as the furnace ages, whereas glass melt contact corrosion (below metal line) may slow down with furnace age.

Figure 13

In addition to studying corrosion behavior, we also measured changes in the physical properties of the AZS samples. In general, the level of apparent porosity was found to be higher than that seen in as-is AZS. This increase is most likely due to the formation of new crystalline phases (which can be higher in density and thus lower in volume) at the expense of the glassy phase, and also from liquid phase rundown from the refractory surface into the glass bath.

Figure 14

Figure 14 explores a correlation between the AZS refractory corrosion and knot/cord-type glass defects. In order to create the chart, we have plotted the chemistry of all knot and cord defects analyzed at the Monofrax Techni-

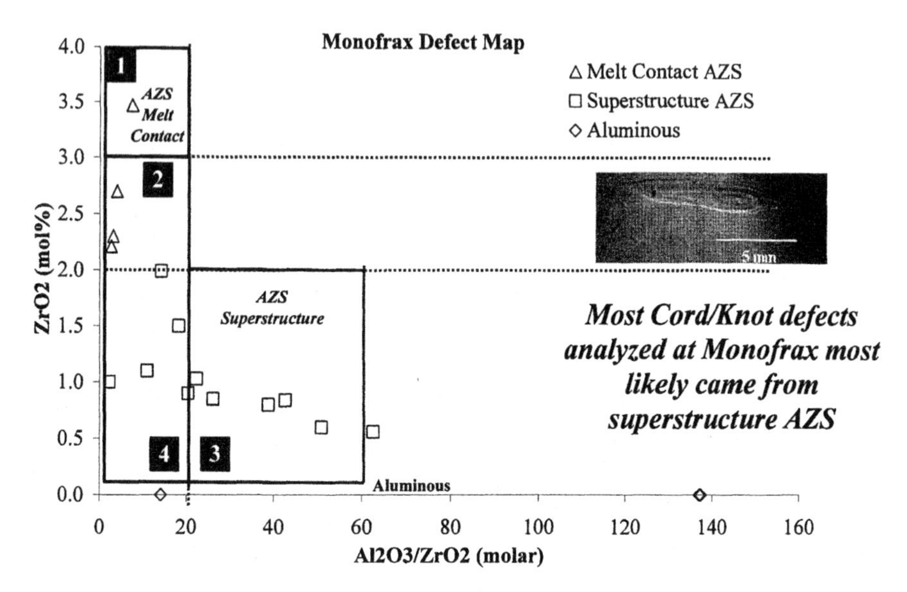

Figure 14. Accurate EDS analyses of the knot/cord center region provide a reliable indication of the defect's source based upon the defect's molar ZrO_2 and Al_2O_3/ZrO_2 ratio.

cal Center in the last 5 years or so. The chart compares the ZrO_2 concentration with the Al_2O_3/ZrO_2 molar ratio of the defects. When assigning the most likely origin of the defect (i.e. superstructure vs. melt contact AZS corrosion), we have made use of the chemistry of the AZS refractory samples discussed in this paper and also many other samples from our database.

The defect represented by a triangle symbol in Area 1 is believed to have definitely originated from melt contact corrosion of AZS refractories. The defects represented by square symbols in the Area 3 are believed to have definitely originated from superstructure corrosion of AZS refractories. Defects covered in Areas 2 and 4 represent some uncertainty about the source, although we believe defects in Area 2 are most likely from an AZS melt contact source, and likewise the defects shown in Area 4 are most likely from an AZS superstructure source.

Because the defects shown in this chart came from furnaces melting many types of glass chemistry, and because the majority of the defects appear to have originated from superstructure AZS corrosion, we conclude that as a given glass-melting furnace ages, glass defect formation does continue.

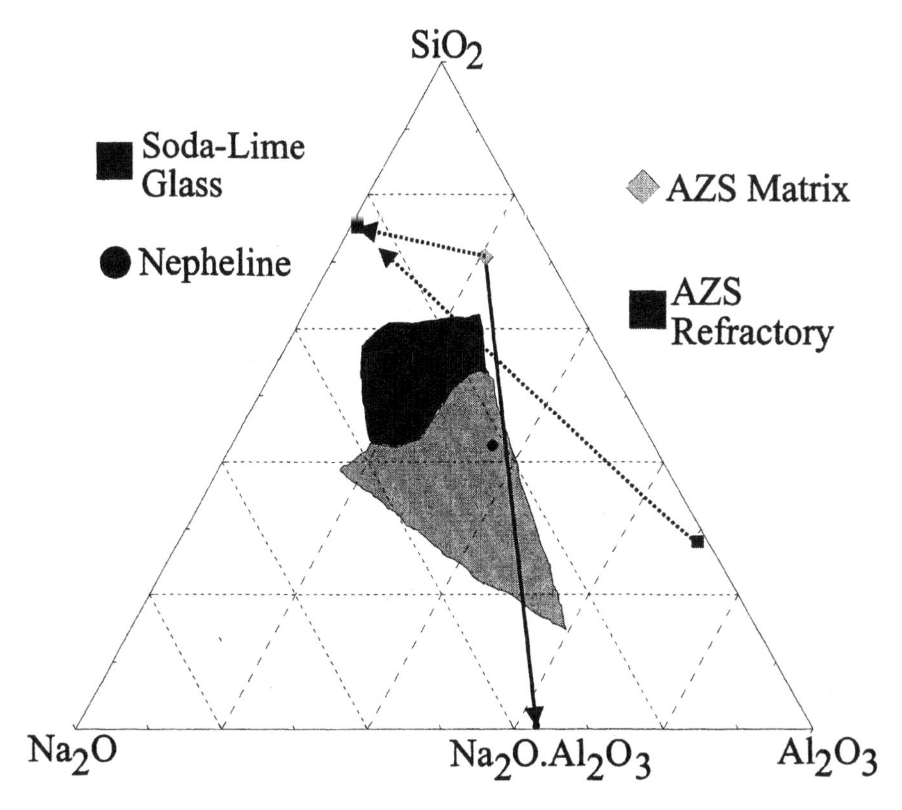

Figure 15. Data on AZS corrosion shown is in agreement with the phase diagram prediction.

Figure 15

The soda/alumina/silica ternary phase diagram from *Phase Diagrams for Ceramists* shown in Fig. 15 has been modified to show the chemistry of an as-is AZS refractory, the chemistry of soda-lime silica glass, and the likely result of the interaction of the AZS with the glass. When AZS refractory is exposed to the soda-lime glass furnace environment, the path of the chemical interaction may be expected to follow the arrows originating from the gray symbols to the black symbol showing the glass composition. However, in reality, for every mole of soda diffusing into the refractory matrix viscous phase, one mole of crystalline alumina is also dissolved in the matrix phase. This drives the matrix chemistry toward the the 1:1 Na_2O/Al_2O_3

point, sending it through the nepheline phase field. This prediction from theory is consistent with the nepheline corrosion products seen in the AZS matrix region after service in soda-lime furnaces.

Summary

The following statements are presented in summary.

1. Both short- and long-term corrosion mechanisms are similar.
2. Glass melt contact refractory corrosion can passivate because of boundary layer formation and the effect of external cooling at the metal line.
3. Superstructure refractory corrosion, however, can continue through the entire campaign duration (recall the results from soda-lime container tank, corrosion depth ~50 mm at glass contact, ~150 mm in the superstructure).
4. Post-campaign refractory evaluation has also shown an increase in apparent porosity, which may be due to formation of new phases and/or matrix (liquid) phase rundown.
5. Knot and cord chemistry appear similar to AZS hot face chemistry following corrosion.
6. Though both glass contact and superstructure corrosion products can lead to knot/cord defects, superstructure corrosion is a more potent and long-term source of defects.
7. Used AZS may contain new crystalline phases (nepheline, kalsilite, leucite, β-alumina, zircon). A mismatch in the coefficient of thermal expansion with unaltered AZS may lead to spallation, causing defects.

Conclusion

As mentioned earlier, the glassy matrix phase in AZS (which is effectively one-third of the total volume) provides big pathways for the corrosive alkaline and alkaline earth species to diffuse into the body of the refractory. This in-diffusion promotes the dissolution of crystalline alumina, resulting in an expansion of the glassy phase volume. The data shown in this paper from post-campaign AZS superstructure blocks show significant chemical alteration of the glassy phase up to several inches of the block thickness.

Given that the majority of the knot and cord defects analyzed at

Monofrax are similar in chemistry to that of the AZS glassy phase following superstructure corrosion, and that AZS superstructure corrosion may progress over the life of the furnace campaign, it is reasonable to conclude that superstructure AZS corrosion is an ongoing source of glass defects. The rate of defect generation is, however, a more complex issue and depends on many other factors besides refractory degradation. These factors include furnace temperature profile, throughput, and furnace exhaust control.

A correlation between glass defect frequency and furnace would require meticulous record-keeping of furnace process conditions and defect level over the furnace campaign. One can either engage in this long-term study or determine better alternatives to reduce defects based on current understanding of refractory degradation. There is no doubt that all AZS refractories experience an expansion of the glassy phase volume due to superstructure corrosion and can therefore serve as a source of liquid phase rundown, which promotes glass defects.

Fusion-cast α/β-alumina refractory contains a very small amount (~2 vol%) of crystalline boundary phase–bearing nepheline-type chemistry. Comparative studies of AZS and α/β-alumina refractories superstructure corrosion in air-fuel and oxy-fuel furnaces have shown significantly lower chemical alteration of the α/β-alumina than that seen in AZS. Furthermore, over the last 10 years, α/β-alumina refractories have been successfully used in the crown and superstructure of oxy-fuel-fired glass melting furnaces, showing excellent physical and chemical stability over multiple campaigns.

Therefore, Monofrax strongly recommends the use of α/β-alumina refractory for use in the superstructure lining in air-fuel furnaces for effective defect minimization.

Bibliography

Aydin, "The Evolution of Knots and Cords in Glass Products"; in *XIV Intl. Congr. on Glass.* 1986.

Beck and Weibmann, "Investigation of Cat Scratches on Container Glass," *Glastech. Ber. Glass Sci. Technol.,* **74** [4] (2001).

Dunkl, "Studies on the Glass and Reaction Phases Given Off by Fused-Cast AZS Blocks and Their Effects on Glass Quality."

Duvierre, Krings, and Sertain, "Defects and Their Origin in Glass," *Glasteknisk Tidskrift,* **45**, 2 (1990).

Heitz, "Identification of Lead Crystal Glass Defects Related to Refractories: Advantages of SEM Microanalysis."

Van Dijk, "Knot Formation Due to Glass Melt/Fusion Cast AZS Interaction."

Review of Improved Silica Crown Refractory and Practices for Oxy-Fuel-Fired Glass Melters

Alonso Gonzalez R.
Grupo Pavisa, Naucalpan, Mexico

John T. Brown
GMIC, Corning, New York

Roger P. Weilacher
Glass Design, Inc., Sapulpa, Oklahoma

Michael A. Nelson
Consultant, Jeffersonville, Indiana

For many years, silica has been the preferred melter crown refractory for air-fuel-fired furnaces. Silica refractory brick are relatively low in cost yet quite serviceable, with their service life usually that envisioned for the glass-melting furnace campaign.

This standard for conventional silica for air-gas melters has been a low flux factor (defined by ASTM as Type A where the amount of Al_2O_3 plus twice the amount of alkali must be 0.50% or less) or super-duty silica brick, with a typical density of 111 to 115 lb/ft^3 and an apparent porosity of 20–24%. These products give good performance (12–15 or more years in many cases) and are relatively inexpensive. We can pick a price of about $750 per ton and assign this a cost factor of 1.

About 10–15 years go, when soda-lime glass furnaces were converted to oxy-gas combustion systems, the life of silica superstructure, particularly the melter crown, became somewhat shorter. Some of this came from direct corrosion of the crown's hot face. Rat-holing of the silica crown was also noted as being much more prevalent with oxy-gas melters. The conclusions reported by many technical investigators can be summarized by stating that the cause was the higher Na^+ and OH^- concentrations in the products of combustion from the oxy-gas flame.

A joint investigation by Harbison-Walker and Corning, Inc., indicated that the calcium silicate bonding phase in conventional silica brick was

Figure 1. Alkali vapor testing. Left: conventional silica, 6.5–7 mm erosion depth, 3–5 mm penetration depth. Right: Low-lime silica, 1.5–2 mm erosion depth, 1.5–2 mm penetration depth.

being significantly attacked by this higher concentration of aggressive species in oxy-gas melters. This team then decided to look for alternative solutions within the silica refractory field.

Their efforts concentrated on three materials. First, as a reference for all comparative testing, was standard silica brick. Second was fused silica, a product whose behavior in oxy-gas environs was known to be better than standard silica. However, fused silica was ruled out as a crown material solution because of significant changes in phase mineralogy. Above 1200°C it will devitrify into crystalline silica, resulting in cristobalite and/or tridymite. This phase change is accompanied by a notable volume reduction.

The third material was actually series of silica compositions with lesser amounts of CaO than standard silica. Laboratory and pilot furnace tests showed that much better performance was realized with a product that had 0.8% CaO and lower apparent porosity when compared to standard silica refractory brick. This material is referred to as low-lime silica.

Samples of conventional silica and low-lime silica were compared via ASTM C-987, an alkali vapor test run at 1370°C for 24 h (Fig. 1). When the samples were split and analyzed, the conventional silica on the left was corroded to a depth of 6.5–7 mm and was penetrated by alkali an additional 3–5 mm. The low-lime silica on the right was corroded to a depth of 1.5–2 mm with alkali penetration of another 1.5–2 mm. Clearly, the low-lime silica body was superior in this test.

Sample pieces of the conventional silica and low-lime silica were tested in the superstructure of an operating oxy-gas-fired soda-lime container furnace. Figure 2 shows post-test elemental maps of the refractory hot face of the two materials after an 85-h exposure at 1540°C. We are looking at a 600

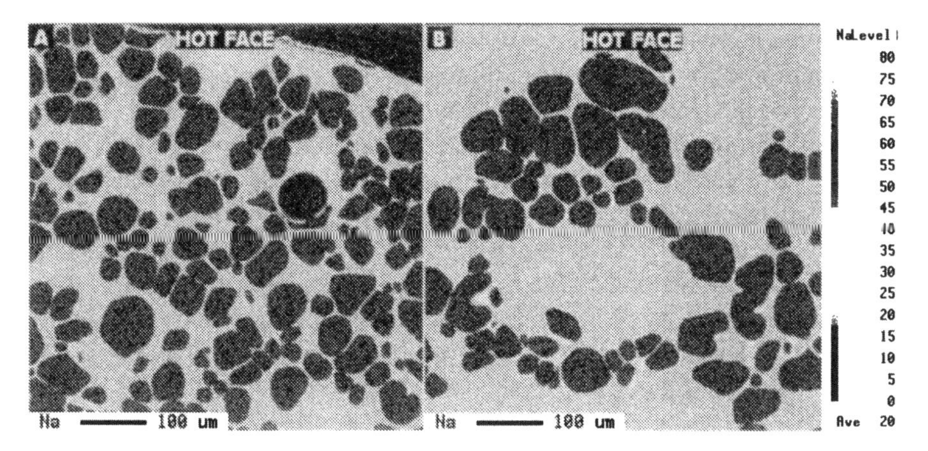

Figure 2. EPMA X-ray map, 85 h at 1540°C, brick hot face. Left: Low-lime silica. Right: commercial silica.

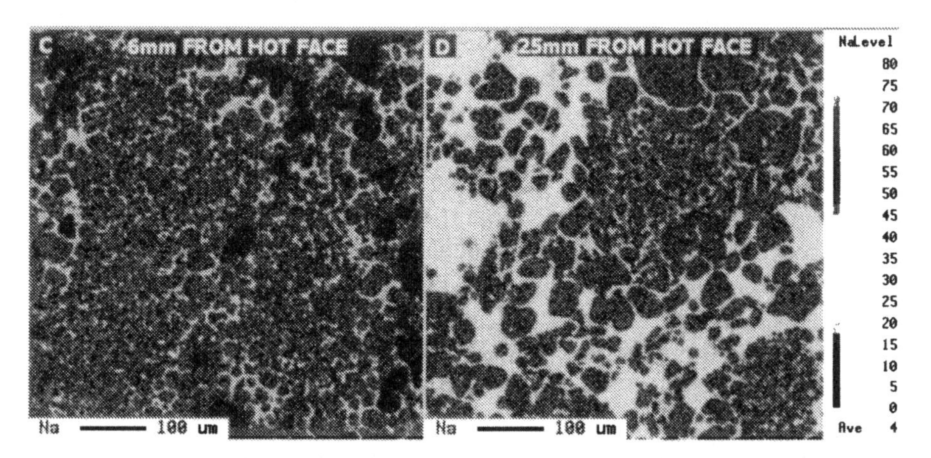

Figure 3. EPMA X-ray map, 85 h at 1540°C, 6 mm from hot face. Left: Low-lime silica. Right: Commercial silica.

× 600 μm section with quantitative elemental analyses done on 1 μm spots. Therefore, each photo represents 360 000 analyses per element. Note the far greater amount of reaction for the conventional silica specimen. At the hot face, low-lime silica clearly has more unaltered silica grains than conventional silica.

Figure 3 looks at the interior of the two test pieces, again at 1540°C for 85 h of exposure in an industrial oxy-gas furnace. We see that at a depth of

25 mm the conventional silica still has a fair amount of dissolved material surrounding the remaining silica grains. This is from penetration of the sodium into the interior of the refractory. For the low-lime silica on the left, we are only 6 mm from the hot face and we see less reaction and soda penetration. This clearly shows that low-lime silica has the potential for superior performance in a melter crown as compared to standard silica.

If we digress a bit, we can take a brief look at some alternative melter crown refractory materials. The first case is for fused refractories as fused alumina and fused AZS. There is now considerable experience with these materials in furnaces melting various glass compositions, including soda-lime. They have a low defect potential (it is lower for fused alumina than for fused AZS) and it appears they will have long life. However, because of their density and the use of large blocks, support steel must be replaced and strengthened. All of these points yield a cost factor for fusion cast crowns of 8–15 compared to conventional silica.

A second alternative group would be bonded brick of AZS, spinel, mullite, and alumina compositions. These have also found specific applications but not particularly in soda-lime glass melters. Although there are always exceptions to the general case, it can be said that their campaign life is not fully known at this time and they likely offer the prospect for higher defect levels than fused cast because of their open or continuous porosity. A bonded crown may require stronger support steel because of its higher density compared to silica brick. However, the cost penalty for most bonded crowns is less than for fused cast. It is thought that they fall in the range of 4–10 times that of conventional silica brick crowns.

With this cursory summary of more expensive crowns behind us, we can return to silica crowns. The goals for silica crowns in oxy-gas melters can be summarized as follows:

1. Peace of mind.
2. Keep crown refractory investment low.
3. Seek typical long silica crown life.
4. Look at construction, heatup, operation.
5. Evaluate “improved” silica products.
6. Make soda-lime glass more competitive.

If we review these goals, our first thought is the peace of mind one would get from a better silica crown system. This is a case where expensive hot repairs are not necessary and costs associated with an interim crown repair during the furnace campaign are possibly eliminated.

Of course we would want to keep increases in refractory investment at a minimum and still get a full, long silica crown life. To do this, we must also look at crown construction practices, crown heatup, and furnace operations, keeping in mind our ultimate goal of making soda-lime glass products more competitive.

To realistically accomplish this, it was necessary to evaluate a "new and improved" silica crown brick product. To that end we should review the development of oxy-gas silica Product X as this new and improved silica composition. Based on Corning laboratory studies, the production target for CaO was ~0.8% rather than the 3+% for standard silica brick. The density for the new product was notably higher at ~117 lb/ft^3. The porosity was significantly lower at ~18%. It is worth noting that Corning patented this work in 2001. Detailed information about this project was published at the Parma meeting in Italy in 1999.

Prior to closing the silica brick manufacturing plant, one production run of this new low-lime silica brick was produced for an order by Grupo Pavisa in Naucalpan, Mexico. Low-lime silica brick from this single production run were installed in a furnace that was enlarged by a factor of two from a previous smaller oxy-gas melter. The details are given in Table I.

The melter crown (Fig. 4), with a 60° central angle, has fused AZS skews and is made with 15-in. thick low-lime silica brick. There is also a 15-in. thick medium-density zircon brick between the skews and the silica brick as a neutral barrier. A positive double drip course of silica was also employed. On top of the silica brick there is a 3-in. layer of Harbison-Walker Crown Seal II (a monolithic fused silica seal material), a 2.5-in. layer of Lubisol-1 castable, a 2.5-in. layer of Lubisol-2 castable, and a 1-in. layer of Lubisol-3 castable. These last layers help to insulate the crown so that the interface temperature between the silica and the sealing material is raised above the condensation temperature of alkali-rich vapors. Also, note that all the insulation is monolithic, preventing penetration of these same aggressive alkali vapors.

The Grupo Pavisa furnace started in late January 2000. Recent inspections, including those by a visiting group in June 2003, indicate that the crown appeared smooth and dry with no rat-holes and no noticeable wear. The silica drip course had sharp edges for the entire length of the crown.

Table I. Grupo Pavisa oxy-gas melter

Melter area	15.5×34 ft = 527 ft^2
Glass depth	50 in.
Pull rate	150 metric t/d
Bottom boost	600 Kwh
Fuel	Natural gas @ ~1000 Btu/ft^3, 100% oxy-gas, on-site O_2 production
Glass	Flint soda-lime container
Cullet ratio	~50%
Melt temperature	1530°C (2786°F)
Burner layout	6 opposed (3 per side)
Fuel use	~3.6 MMBtu/t, 200 Kwh boost, ~135 M t/day (5/13/03)

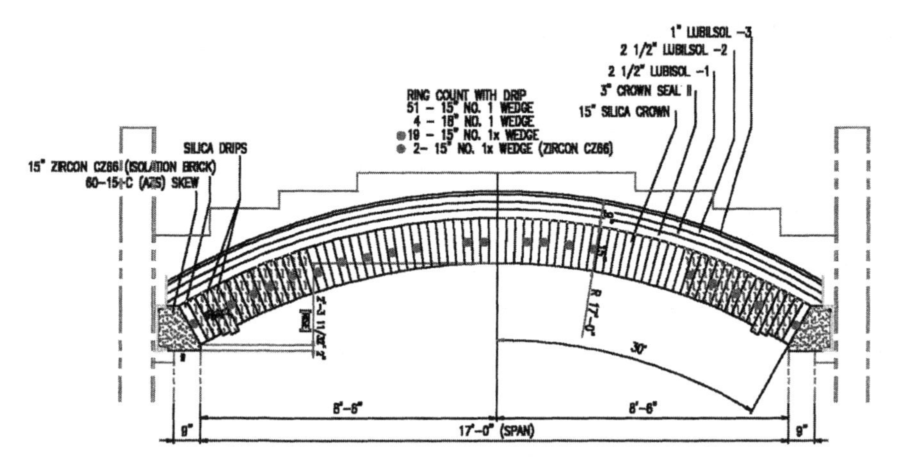

Figure 4. Melter crown drawing.

Also, each brick joint in the crown was well defined and did not suffer from the wash of corrosion and/or erosion. The only wetness observed was an approximately 1-m semicircular area over each batch charger.

The glassmaker's experience can be summarized as "happy and satisfied" with the crown at this point in time. It is worth noting that he took some risk in installing a new brick but that nothing bad happened and there have been no instances of deleterious performance by these bricks.

Based on the information to date, it can be said that this glassmaker would, in the future, pay some amount of premium for this type of solution to oxy-gas melter crown problems normally expected with conventional silica in oxy-gas melters. For 10 years they have been melting container glass with oxy-gas combustion systems in eight different melters. The glassmaker wanted to purchase two additional crowns but the existing supplier was unable to find a new supply facility to provide the low-lime silica brick.

We now leave the area of material composition and look at the physical specifications for silica crown brick. Once the correct silica brick chemistry is secured, a key consideration may be the dimensional tolerances and the flatness of silica brick. Consistent brick sizing along with flat faces will minimize the amount of mortar required. Thick mortar joints become weak points in the crown that provide sites for attack by aggressive species in the combustion products. A target of 0.03125-in. joints is suggested, with an acceptable maximum of 0.0625-in. joints. Machining of some faces may be an option worth considering.

When the silica crown bricks are sorted and grouped by size, insist on each group having a very small size range. This will allow very little variance in the thickness of an individual row during crown brick installation, thereby yielding a tighter fit of the brick and smaller mortar joints. Using oversized brick or jumbo sizes will reduce the number of joints, which is also a step in the right direction for fewer crown problems.

Proper construction of the furnace crown is another key element in securing a long and useful melter life. Prior to construction of the silica crown, all skew channels must be level, plumb, and square. This facilitates skew installation and results in the skews also being level, plumb, and square. They provide a good and proper base for the installation of the silica crown bricks.

All furnace crown silica brick construction starts with the support forms and how they are installed. All centers need to have a true and even radius. The deadwood, or straight, at the end of the center must be the same for each and every center. When the centers are being fabricated, a single pattern should be used to ensure that all centers are the same.

The shoring should be adjustable so that lagging can be set to the crown skew. When installing lagging, keep the strips in a straight line. After the installation of the lagging, make sure that no high or low spots are present.

After the complete center has been installed, the next two important steps to be considered are the crown layout and the stocking of refractories. When the lagging task is completed, the furnace centerline must be established and chalk lined on the lagging. Parallel chalk lines then should be applied to the remainder of the lagging.

After the crown layout has been completed, stocking of refractories can begin. Conveyor lines need to be set up so the brick do not bang together. This will eliminate new corner spalls and edge spalls. When stacking brick on the top of the crown center, make sure that the different brick shapes are kept in separate piles. When handling the brick, care should be taken that brick are not dropped or banged together, which causes chips and spalls.

Once the crown is stocked with brick, laying of the silica brick is ready to start. As always, it is extremely important to have qualified craftsmen doing the brick installation. The crown should be installed on day shift only, using the same crew until the crown is completed. Accomplish and constantly maintain proper silica mortar consistency at all times during the installation of the crown brick. Always check to make sure that both sides of the crown are parallel.

Use the minimum amount of silica mortar and keep the head joints and bed joints tight. Thick mortar joints are a weak area prone to chemical attack.

Tap the brick into place but do not over-hammer them while installing the melter crown. Use no small cuts for closure key brick. After the centers have been removed, carefully wash and tuckpoint the hot face of the crown.

When reviewing melter crown construction, we can summarize the steps by emphasizing the four key basic areas as crown center fabrication, crown center installation, stocking of refractories, and installation of the refractory material.

When we look at the total silica refractory crown for oxy-gas fired glass melters, it may be thought of as a three-legged stool where each leg is critical to the stability of the stool, or system.

The first leg is the silica refractory itself. It must be a product of the proper chemistry and/or manufacturing process for the crown brick. Coupled with this are dimensional control of the brick and flatness of the brick faces.

The second leg is the actual installation of the silica crown brick. Shortcuts to save time here, usually near the end of a repair, can offset any gains made from selecting premium silica brick.

The third leg is not addressed by this paper but is equally important. If the melter heatup is not handled correctly, efforts made toward improved silica refractory selection and brick installation can be undone. The same holds for insulation design, operation of the furnace, and inattention to hot repairs or maintenance procedures.

In summarizing this presentation, our goal has been to allow the industry to examine a total system and outlook when looking to silica brick for long melter crown life with oxy-gas melters. We offer ten points to consider.

1. We must first look for the optimum brick composition and its accompanying manufacturing process. In short, it is not just a matter of decreasing the CaO level but also making changes in production processes that allow marketing of an improved product.
2. The manufactured brick must hold to minimum dimensional and flatness specifications to allow its installation with a minimum of mortar.
3. We have presented steps and actions for consideration relative to achieving proper installation of the silica brick. A premium material that is incorrectly installed will likely not perform well.
4. For the case at hand with the low-lime silica brick, we have what appear to be good results to date. The downside is that this is the only installation of this particular composition. Not only will this crown continue to be closely monitored, but so will oxy-gas melter crowns built with other oxy-silica brick products. Successes and failures should be noted, along with the reasons behind them.
5. There has been interest in exploring the licensing of the patent Corning, Inc., holds for the low-calcia silica brick composition. Currently, a domestic manufacturer of silica is in preliminary discussions with Corning.
6. We must continue to collect input regarding the subjects noted for our three-legged stool. Anyone who can contribute their experience or observations regarding these subjects should do so. Here we have not covered furnace heatup, operation, or hot repairs as specifically related to silica crowns in oxy-gas-fired soda-lime tanks. If new and effective information is out there, we ask that you bring it forward.
7. The improved composition of silica brick, noted earlier, along with tighter dimensional and flatness specifications, will likely increase the cost of the melter crown. It has been estimated that this could be 1.5 to 3 times the cost of conventional silica brick.
8. Remember that manufacturers of silica brick and silica products must be profitable in order to continue serving our industry.

9. We ask for more papers and/or publications regarding your experience or learning curve with silica crowns on oxy-gas furnaces.
10. To solve our current problems and to make glass a more competitive product, we must all share our facts and our technical knowledge, publicly or privately, regarding silica crowns for oxy-gas melters.

Bibliography

Understanding and attacking the multiple conditions that contribute to the destruction of silica crowns in these corrosive conditions is an accumulation of knowledge contributed by many others, including the following.

Mark D. Allendorf et al., "Analytical Models for High-Temperature Corrosion of Silica Refractories in Glass-Melting Furnaces," *Glass Sci. Technol.,* **76** [3] 136–151 (2003).

Mark Allendorf and Karl Spear, "Thermodynamic Analysis of Silica Refractory Corrosion in Glass-Melting Furnaces," *J. Electrochem. Soc.,* **148**, B59–B67 (2001).

Tom Clayton, "Protecting Silica Refractories in Oxygen Furnaces," *Glass,* July 2001, pp. 186–187.

Harbison-Walker Modern Refractory Practice, Fifth ed., 1992.

John LeBlanc, "Controlling Silica Attack on Soda-Lime Oxy-Fuel Furnaces," *Ceram. Ind.,* June 1996, pp. 27–29.

M. Velez et al., "Evaluation of Crown Refractories under Oxy-Fuel Environments"; presented at the 7th International Conference on Advances in Fusion and Processing of Glass, 27–31 July, Rochester, New York.

D. S. Whittemore, R. F. Spaulding, H. E. Wolfe, and J. T. Brown, "New Silica Refractory for Oxy/Fuel Glass Melting," *J. Int. Glass,* **102**, 120–124 (1999).

Engineered FKS Platinum Solutions for High-Temperature Applications in Today's Glass Production

Michael Oechsle, Hubertus Gölitzer, and Rudolf Singer
Umicore AG & Co. KG, Hanau-Wolfgang, Germany

Fine-grain-stabilized platinum and its alloys have a successful history of use in long-term applications in high-temperature glass-forming processes. The unique properties of grain-stabilized platinum materials offer significant advantages when combined with optimized forming and manufacturing techniques. This paper outlines an approach involving the use of advanced simulation techniques that, when combined with detailed engineering and fabrication processes, offers cost-effective solutions for the glassmaker and produces a high-quality product in elevated temperature environments.

Introduction

The term "FKS platinum" (fine-grain-stabilized platinum) is increasingly becoming a synonym for high-grade platinum components for the glass industry. The primary reason for this development is that FKS platinum construction materials possess unique mechanical properties at temperatures that approach the melting point of the metallic matrix. This technological advantage over pure platinum is based on the fact that it is possible to adjust the grain structure of FKS platinum as desired, which makes it possible to increase a material's strength. The excellent thermochemical resistance of platinum alloys compared to glass melts permits the manufacture of corrosion-resistant components. Potential glass defects caused by materials coming into contact with the glass can be reduced so that glass products of the highest quality can be produced.

Because factors other than the material properties alone are also critical to the satisfactory operation of the glass tank, every component placed into the high-temperature domain is subject to an expensive design process. Typically, redundancy is built into complex manufacturing systems to ensure a fail-safe process. To determine the solution concept, it is necessary to answer interdisciplinary problem definitions from the field of glass/metal to take technical aspects into consideration, and to incorporate side constraints and geometric requirements.

For the specialists of Umicore, the functional interlinking of extremely different special disciplines is one of the greatest challenges posed by the development process. In addition to empirical values determined by practitioners, results from computer simulation calculations are gaining in importance. In recent years, life cycle management was consequently established as the principal tool for bringing together the different areas of expertise for structural components made of FKS platinum materials.

Approach

The redundant design of components made of FKS platinum materials requires an understanding of the interaction of different technological disciplines. Thanks to the use of modern computer-supported development tools, these physical laws can be connected through mathematical relationships. Here the primary goal is to represent processes and procedures visually in time-discrete and spatial form. Variation of parameters then permits us to determine the dominant influencing variables and, from them, to derive spheres of work for later industrial use. These placative results allow us to optimize known engineering parameters, such as the choice of material, structure, and forming.

Side Constraints on the Process

The glass-processing industry subjects components made of FKS platinum materials to extremely different external and internal stresses during industrial use. We will now discuss some of these relevant stresses.

Heat

Among other things, an understanding of thermal balance is critical for controlled temperature control during glass processing. To a first approximation, thermal balance is based on two basic principles in thermodynamics: heat transmission and heat conduction.

The equation of heat transmission $Q = \alpha \cdot A \cdot t \cdot \Delta T$ is based on the fact that liquid bodies that come into contact with a solid body of another temperature will either give off or receive heat from this other body. In heat conduction, thermal energy is passed on within a body. Stationary heat conduction describes a state of constant temperature. This means that the energy supplied equals the energy carried off. The temperature distribution within the system remains constant. We are then dealing with a constant heat flow $\phi = \dot{Q} = Q / t$, where $Q = (\lambda \cdot A \cdot t \cdot \Delta T) / l$.

If the thermal balance along the glass flow does not correspond to the technical requirements needed for good glass quality, then metallic platinum offers a further advantage. Because of its good resistivity, platinum can be used as an electrical conductor to introduce energy. Direct electrically heated components (Fig. 1) currently represent the state of the art in the manufacture of high-grade technical glasses.

Flow

The total pressure within a flowing fluid is equal to the sum of the static and dynamic pressures and is dependent on the fluid's velocity. Of course, this ideal state does not correspond to reality in systems that involve glass flow. Agitators and other components that interrupt the flow may be used to introduce selective time-dependent and position-dependent imperfections into the melt. It is thus possible to influence the quality of the glass purposefully. Moreover, the glass level and viscosity of the glass affect the flow and therefore the pressure relationships.

Today, computer-supported real-time calculations make it easier to understand these very complex three-dimensional, time-discrete interactions, which are important the design of the flow field. Here, visual representation of the process has proven most helpful of all. Other advantages of flow calculation are the possibility of making a comparative evaluation of the homogenization process, and the possibility of directly transferring spatial pressure fields to the component structure (Fig. 2).

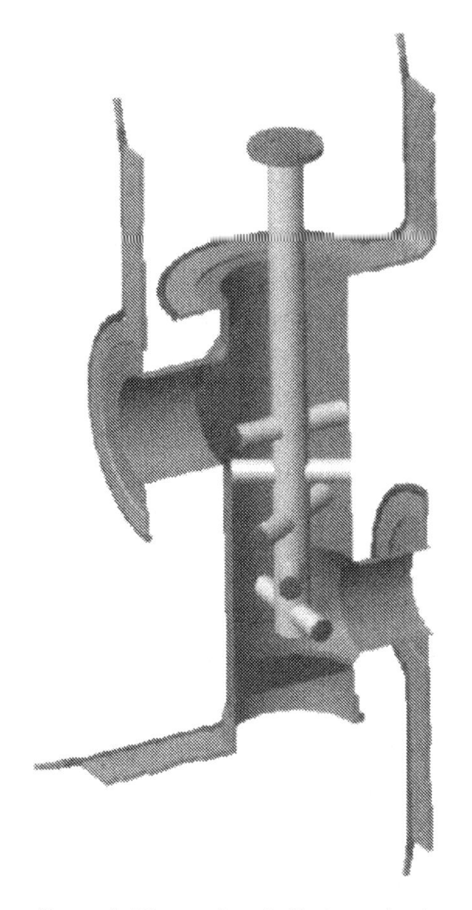

Figure 1. Direct electrically heated agitator cell with bar agitator.

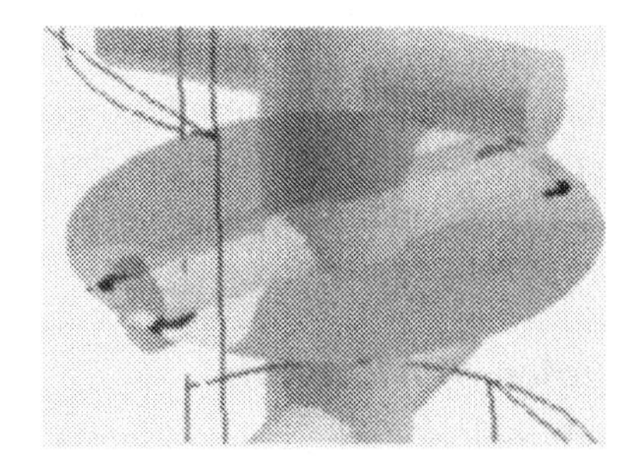

Figure 2. Pressure distribution in a glass melt as a result of agitation effect and glass flow.

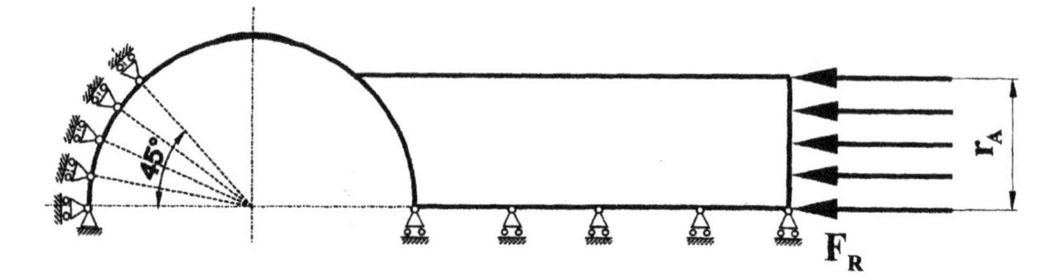

Figure 3. Geometrical storage conditions of a pipe intersection.

Storage Conditions / Installation

As a result of the theoretically given geometric installation conditions, FKS platinum components are subject to deformations and distortions when used (Fig. 3). Detailed knowledge of the properties of deformable bodies is therefore of great practical importance to the operators of glass tanks.

Equilibrium conditions can be of use in examining the static behavior of individual components and subcomponents to a first approximation. It is not possible to draw conclusions about the stress on components on the basis of cutting forces and storage alone. Instead, it is the kinetic relationships that must be considered when solving the typical engineering problems. For this, we can represent the physical regularities between distortion, displacement, and stress in the form of mathematical equations, the stress-strain relation. Three types of equations, which are quite different based on their origin, are used for all of these problems in elastostatics: equilibrium conditions, kinetic relationships, and laws of elasticity.[1]

If we increase the number of undefined side constraints (in relation to practice), the complexity of the systems of equations will also increase. For a first approximation, it is therefore common to make simplifying assumptions that enable us to derive a mathematical solution for the behavior. Computer-supported simulation is a time-saving tool for solving systems of complex simultaneous equations, like those that describe flow behavior. FEM tools make it possible to predict both the static and time-discrete component behavior.[2]

Choice of Materials

High-temperature application places very high requirements on a material's thermochemical resistance as well as on its mechanical properties. Pure platinum fails to meet these requirements from the very start. Platinum pro-

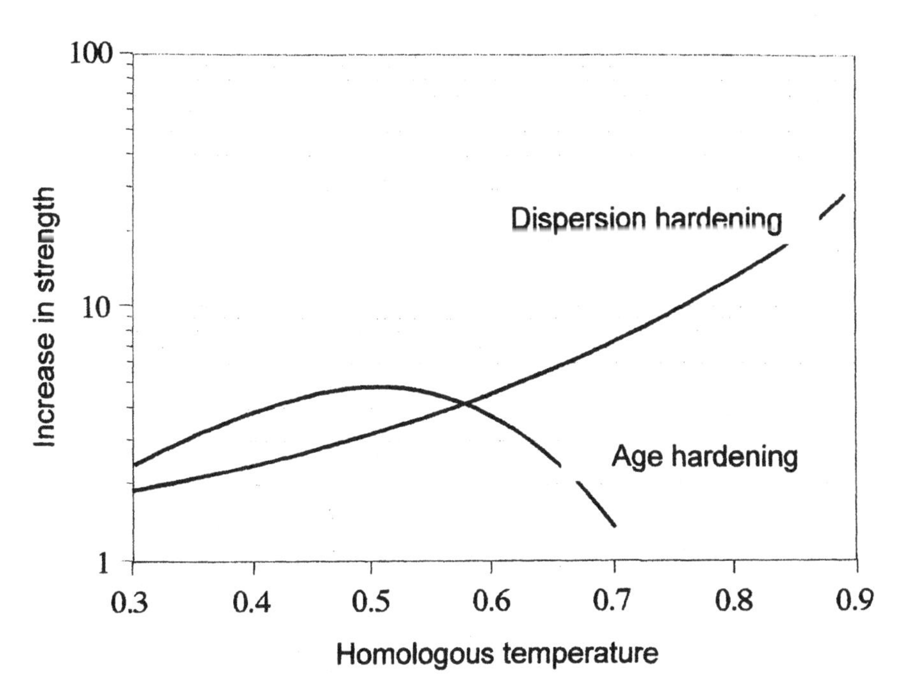

Figure 4. Comparison of the increase in strength due to age hardening and dispersion hardening as a function of temperature.

ducers have consequently introduced procedures to improve the strength of alloys. Material developers have six different mechanisms at their disposal: cold forming, age hardening, fine-grain hardening, texture hardening, solid-solution hardening, and dispersion hardening.[3]

If the thermomechanical stress of the material is close to its melting point, then dispersion hardening is the preferred method of increasing strength. The increase in strength is maintained up to the highest temperatures (Fig. 4). If we now compare the time-dependent and temperature-dependent characteristic values, such as stress rupture strength (Fig. 5), creep behavior, and aging, for typical industrial platinum alloys, it becomes readily apparent how well dispersoids work.

Efficient stabilization of the grain structure in FKS platinum materials depends on the interaction between the oxide particles and the grain boundaries: the boundary surface energy of a matrix particle is reduced as long as a particle remains on a grain boundary. This is true in principle and does not depend on the type of particles.[4] The lower grain surface energy due to

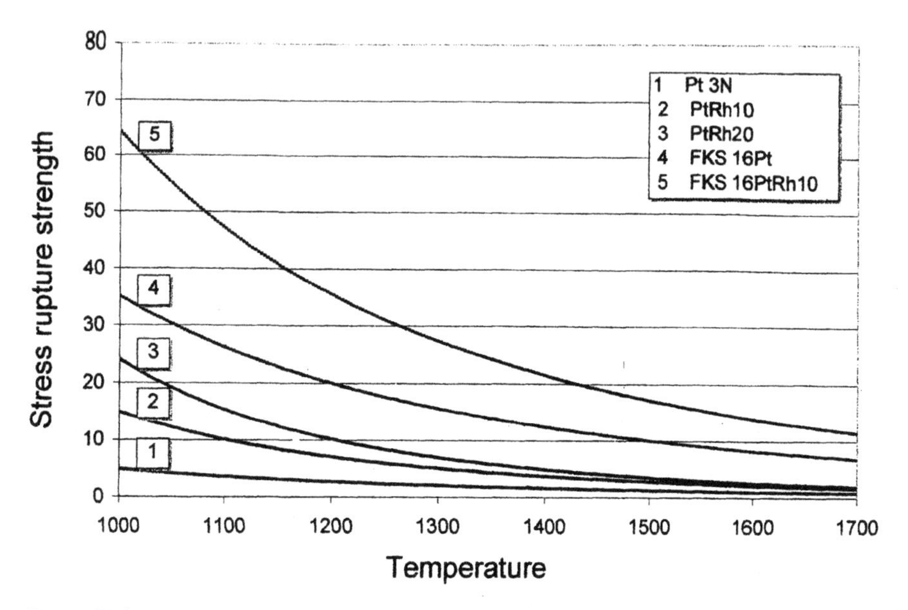

Figure 5. Creep strength of platinum alloys as a function of temperature.

the presence of the particle means that a moving grain boundary can separate from the particle only under a high shear stress. As a result, the grain boundary "hangs together" between two particles (Fig. 6)[5] — or, to put it simply: long-term stable characteristic values for high-temperature application.

Structure

Application-oriented structure is another major reason for the dependability of platinum components in the glass-processing industry. First, it is self-evident that using the minimum possible amount of precious metals is the logical application of the most important principle of lightweight construction. Second, by taking specially developed forming process for the design into consideration, it is possible to create reliable and inexpensive structures.

Lightweight Construction

The principle of uniform utilization of material throughout the entire volume is the basis of the lightweight construction concept. One simple, basic rule is as follows: Move the material from zones of low stress to zones of higher stress, from "inside" to "outside" for most types of stress.

Closed hollow cross sections thus possess a much higher resistance to

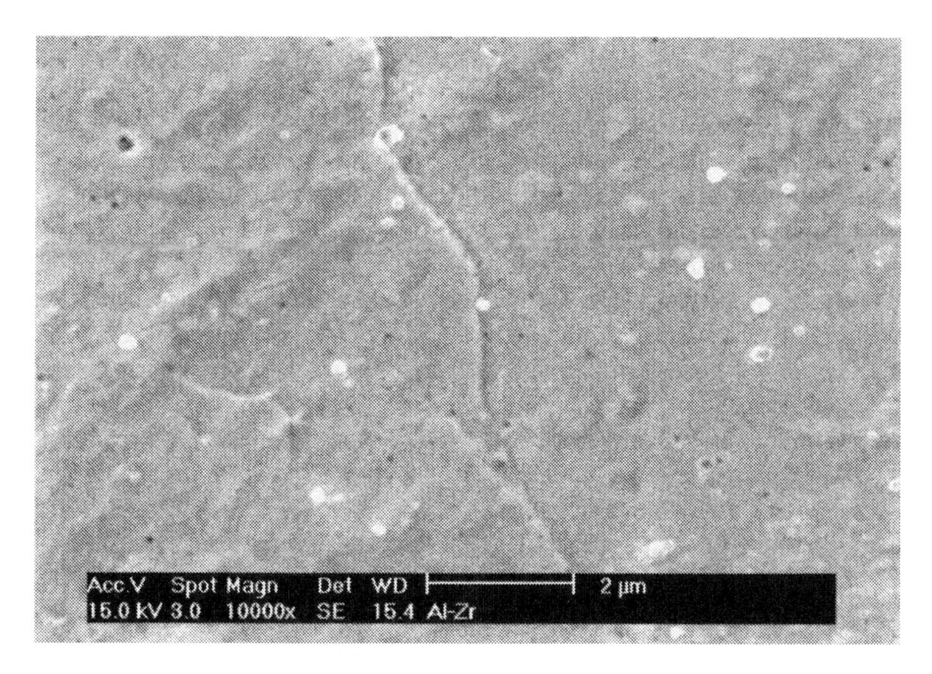

Figure 6. Grain boundary in FKS 16 Pt; the ZrO_2 particles hold on to the grain boundary locally.

torsion than open cross sections. If you compare identical load-carrying cross sectional surfaces, an annular (ring-shaped) cross section with a diameter of 51 mm and a wall thickness of 2 mm for example, then a closed profile can transmit 75 times more torsional moment than a slotted profile. On the other hand, a closed circular cross section would have a diameter of only 28 mm for the same cross-section relationships and would be able to transmit only about 30% of the torsional moment of the closed annular cross section.

Inherent geometrical stability represents another possibility for reducing the use of precious metal. Simple design changes can bring about noteworthy improvements. Consider a conventional smooth pipe. It possesses very high stiffness in the preferred (axial) direction. But this case rarely occurs for a pipe cross section that has something flowing through it. In this case, it is the radial stiffness that must be increased. If we now compare the conventional smooth pipe with a corrugated cross section, then the corrugated form can withstand six times as much external pressure for corresponding geometric relationships.[6]

Forming

The forming process brings together the theoretical prerequisites comprising side constraints, material, and lightweight construction. Here the required component geometry is built from the four basic geometries of FKS platinum: sheet, strip, pipe, and profile. If necessary, smaller component parts are integrally combined into larger components through the selective use of various assembly processes.

Forming is performed in accordance with predetermined regularities. It is based on characteristic values derived from experimental studies. Statistical support for these forming parameters permits us to describe the present problem with technically sufficient precision. The desired form can be created from the basic geometries.

As in the case of material selection, the choice of the procedure that is used is critical to eliminating the need for redundant component use. In addition to the basic material properties, such as isotropy, aging, and formability, the stress that occurs during use must also be considered when establishing the procedure.

Figure 7 shows a pipe cross section produced by hydraulic forming. Hydraulic forming makes it possible to produce thin-walled hollow cross sections with constant wall thickness. Material tapering, like that which occurs during deep drawing, does not occur during the forming process. In addition, the corrugated structure meets the requirements of lightweight construction, namely the highest possible resistance to pressure differences between interior and exterior.

Result

We will demonstrate the method of designing FKS platinum components for the glass industry using a direct electrically heated agitator cell with bar agitator as an example (Fig. 1). Many variants of this typical design are currently in use to improve product quality in the processing of technical glasses.

Design of the agitator cell begins by observing the appropriate agitation process for glass homogeneity. The efficiency of the agitation effect can be derived from the flow simulation. Figure 8 is a plot of two different variants that have identical geometry for the bar agitator. The change in the homogenization behavior is easy to recognize. By taking both the actual state and practical operating experience into consideration, the agitation process can be optimized in accordance with the requirements. Another result obtained

Figure 7. Corrugated pipe made of FKS platinum.

from the flow simulation is the mechanical load on the structural component surface. From this, the designer can derive, most importantly, the mechanical stresses, local displacements, and required driving torque. He will learn whether the characteristic quantities alternately appear as static or dynamic (Fig. 9). In combination with the known temperature-dependent characteristics of platinum materials (namely, creep, aging, and strength), the designer now has enough information to generate the optimum process-controlled forming while minimizing the precious metal weight.

We will use agitator shafts to illustrate the logical implementation of Umicore's lightweight construction concept. The state of the art in the manufacture of agitator shafts is to bend metal sheets round, give them a welding seam, and then work harden them. This concept has several disadvantages. First, the effect of work hardening declines at the operating tempera-

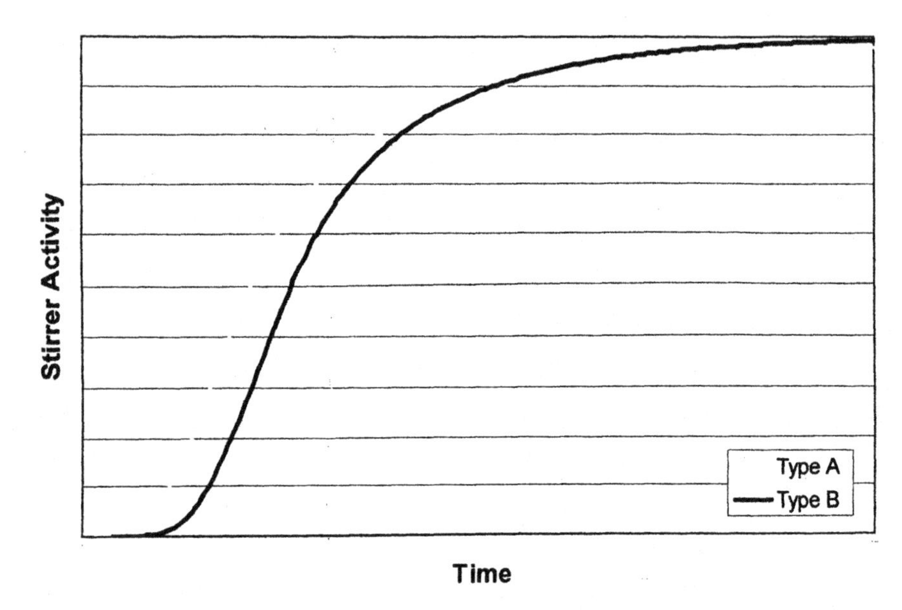

Figure 8. Time-discrete concentration curve as a function of active agitation surface.

tures used. Second, for FKS platinum the joint represents a weak point in the closed profile structure. The apparently closed rigid structure can be stressed only conditionally. If a process requires a high torque at high temperatures, then a large cross section with thick walls is necessary. In this solution, only very large dimensioning can prevent the need for a high amount of precious metal. Of course, these preconditions are not acceptable to the user. Umicore has consequently worked out a solution for this problem definition — namely, the seamless shaft.

The described agitator cell can possess a negative thermal balance as a result of its specified design. The cell is open to the atmosphere, and thermal conduction is directed across the shaft and lining. These two features of the cell's construction cause energy to be removed from the glass melt. If this loss of heat exceeds a threshold limit, then the temperature gradient within the cell will be too high. This could lead to greater problems in the further processing of the glass. To some extent, direct electrical heating of the platinum can help counteract the heat loss and the associated large temperature gradients. This is accomplished by using water-cooled flanges to introduce electrical energy into the metallic material. The electrical energy is converted to heat as a function of the resistivity and is lost to the viscous

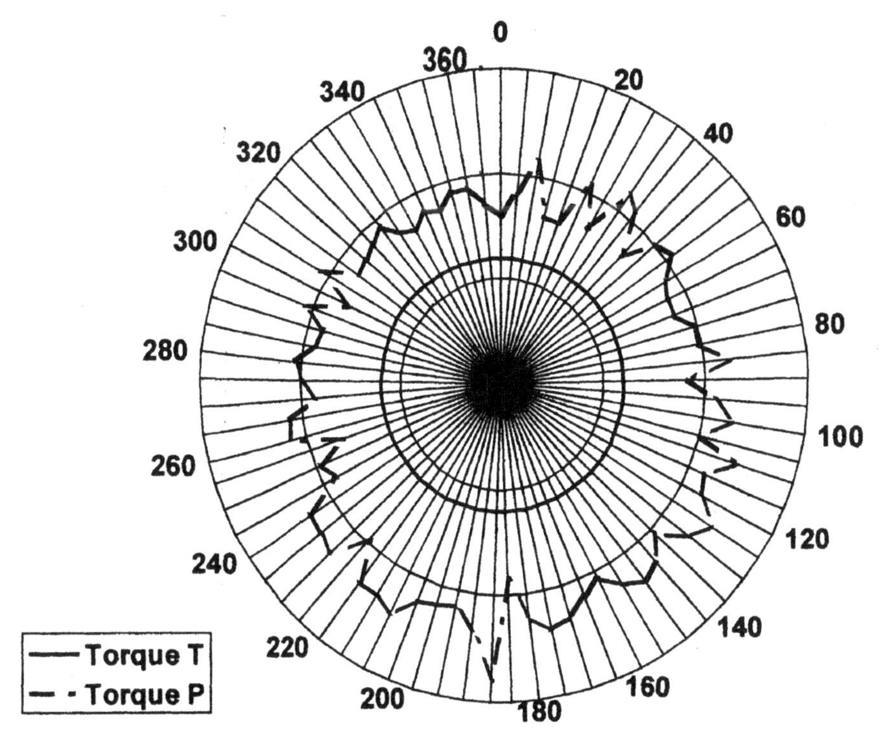

Figure 9. Torque of agitators as a function of dynamic constraints.

glass. The temperature gradient can be selectively influenced within certain limits.

The pipe intersection represents a further potential optimization. Here it is first necessary to answer the basic question: stiff construction or soft construction? In the following, we will use a pipe branch to show the possibilities for a stiffer construction (Fig. 10). This solution is based on the requirement that the spatial position of the system axes (pipe axes) be fixed in all functional stages,

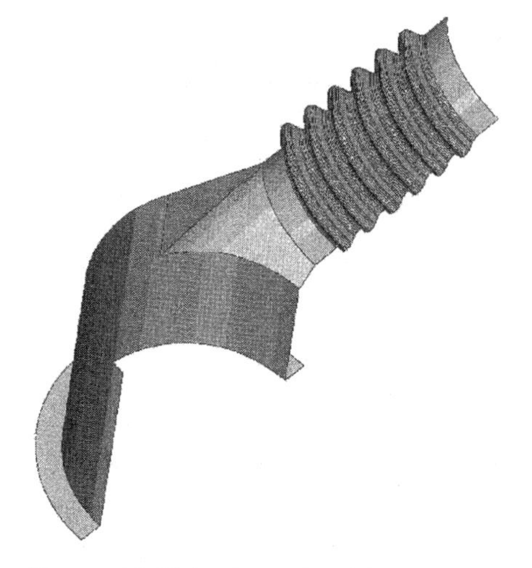

Figure 10. Tube branch with corrugations.

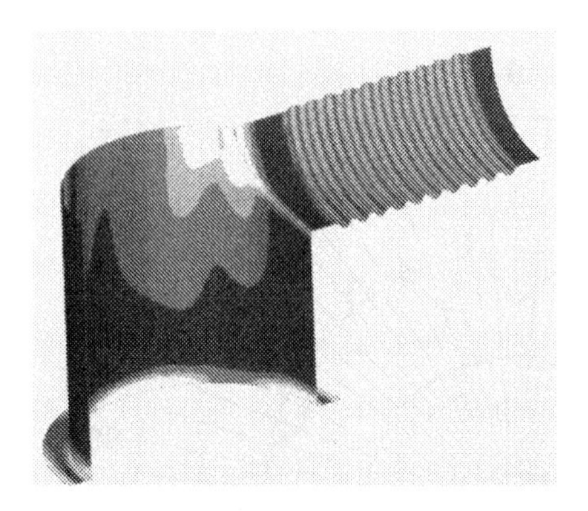

Figure 11. Von Mises stresses on a tube intersection with corrugations.

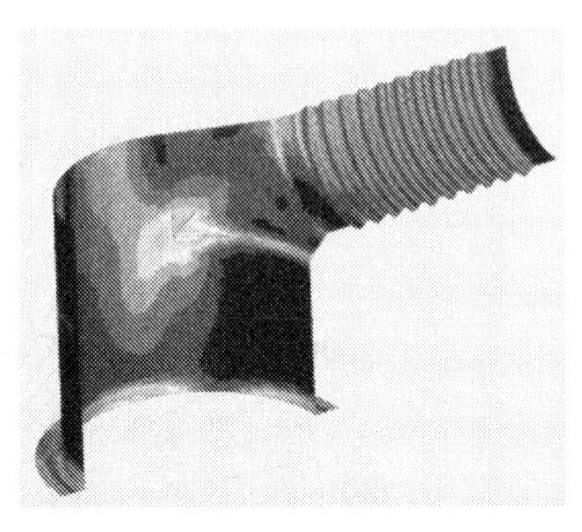

Figure 12. Von Mises stresses on a tube intersection with corrugations.

construction, heating, and utilization. The membrane structure of the cell has to accept distortions and deformations, compensate for them, and diminish them. A pipe branch offers several advantages in comparison to a conventional intersection of two smooth pipes (small inlet, large cell). First, the weak joint can be moved out of the high-stress region. Second, the pipe branch has a soft transition from the inlet toward the cell.

For fixed flanges, like those in the illustrated example (Fig. 1), displacements of the inlet caused by thermal expansion result in a larger distortion within the pipe penetration. In the pipe-branch approach, the entire displacement operates on a much larger surface. This can be seen pictorially in terms of incremental areas. Whereas few increments can be seen along the line of high stress in the case of pipe penetration (Fig. 11), in the case of pipe branch the stress is distributed among many more increments because of the "softer" transition (Fig. 12). The stress per increment could therefore be significantly reduced in the present case.

Conclusion

Functional solutions consisting of FKS platinum materials for high-temperature use have reached a very high state of development. The interlinking of different technological disciplines and the outstanding material properties of dispersoid-strengthened FKS platinum have made FKS a firmly established term in the glass processing industry.

But the user is interested in more than just the redundant method of operation of the components within the process. He also wants the components to have a long service life. In other words, the more efficient and long-lasting they are, the more economical they are. Umicore's life-cycle management has contributed to this continuous improvement process. Umicore's results can be seen in the long-standing, reliable industrial use of key components in glass tanks. This underscores the successful development work of past years.

References

1. M. Rettenmayr, M. Oechsle, S. Zeuner, and W. Krebs, "Eigenschaften von dispersionsverfestigten Platinwerkstoffen," *Glas-Ingenieur,* **3** [13] 2003.
2. Petri Thum, "Steifigkeit und Verformung von Kastenquerschnitten," *VDI Forschungsbericht,* **409**.
3. J. W. Martin and R. D. Doherty, *Stability of Microstructures.* Cambridge University Press, Cambridge, U.K., 1976.
4. M. Rettenmayr, X. Song, M. Oechsle, and S. Zeuner, "Eigenschaften von dispersionsverfestigten Platinwerkstoffen," *Metall,* **56** (2002).
5. W. Schnell, D. Gross, and W. Hauger, *Technische Mechanik Band 2: Elastostatik.* 2nd ed. Springer-Verlag, 1989.
6. M. Oechsle, S. Zeuner, A. Ziegler, and M. Grieb, "Auslegung thermomechanisch belasteter Platinbauteile," *Glas-Ingenieur,* **4** [12] 2002.

Geopolymer Refractories for the Glass Manufacturing Industry

Waltraud M. Kriven, Jonathan Bell, and Matthew Gordon
Department of Materials Science and Engineering, University of Illinois at Urbana-Champaign, Urbana, Illinois

This work briefly introduces geopolymers and some geopolymer composites having typical molar compositions of $SiO_2/Al_2O_3 = 4$, $M_2O/SiO_2 = 0.3$, and $H_2O/M_2O = 11$, where M = Na, K, or Cs. As seen by SEM and TEM, the microstructure was a nanoparticulate (<10 nm) and nanoporous material having a nominal composition of $4SiO_2{\cdot}Al_2O_3$, cross-linked with cations from Na_2O, K_2O, or Cs_2O. Some structural geopolymer composites reinforced with stainless steel powder, basalt chopped fibers, and/or fiber weaves were prepared. Using six layers of natural basalt fiber weave reinforcements, the three-point bending strength and work of fracture of 1 in.$^2 \times 6$ in. long bend bars were increased from an average of 2.75 to 10.25 MPa and from 0.05 to 21.82 kJ/m^2, respectively.

Introduction

Geopolymers, or polysialates, are a class of ceramic materials that can be made at ambient temperatures and pressures. The material is a cross-linked, inorganic polymer made from sources containing both tetrahedral aluminate (AlO_4^-) and tetrahedral silicate (SiO_4) species, under highly alkaline conditions. Such sources can be an aluminosilicate mineral, for example, kaolinite, mullite, andalusite, smectite, bentonite, or feldspar. The mechanism of geopolymerization is thought to occur by three steps:

1. Dissolution of the aluminosilicate in highly caustic solution.
2. Transportation via a water-assisted mechanism.
3. Polycondensation into an amorphous, cross-linked, three-dimensional structure.

Figure 1 is a schematic free energy diagram showing geopolymer formation in relation to zeolites and chemically identical, dehydrated, crystallized ceramics. It shows that polysialates may be thought of as a metastable form of zeolites, which have a lower water content; hence the system lacks diffusion and the ability to crystallize. The lowest energy state in the absence of water, but resulting from high-temperature heat treatment, would be crystalline end products such as nepheline, leucite, kalsilite, or pollucite, depending on the chemical composition of the cross-linking, alkaline species (Na^+, K^+, or Cs^+, respectively).

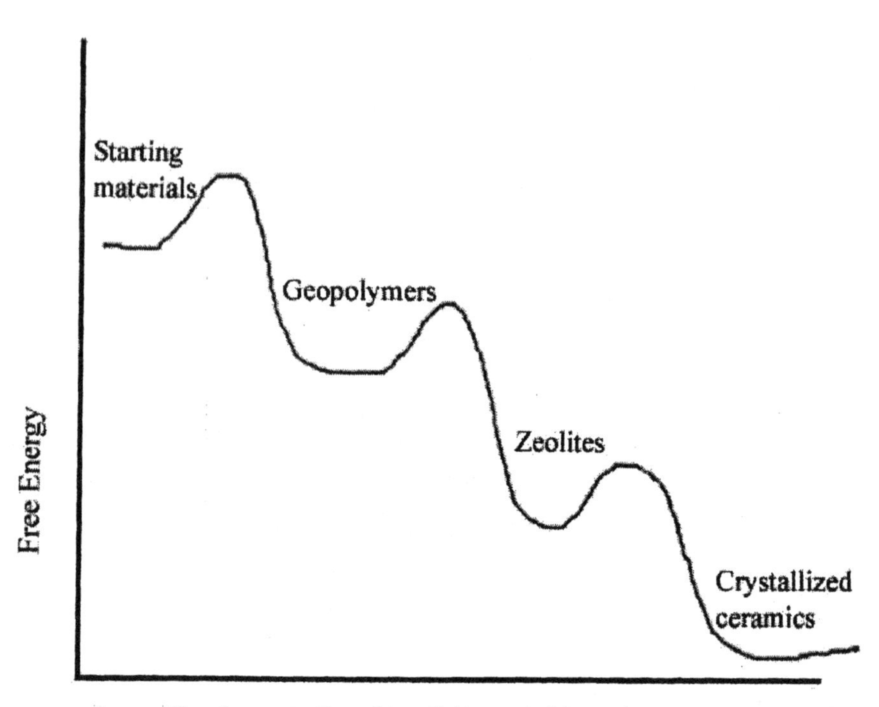

Figure 1. Schematic free energy diagram illustrating the relationship between the metastable and stable phases of chemically related aluminosilicates. Amorphous geopolymers are produced in low–water content mixtures and crystallized zeolites form in the present of plentiful amounts of water, allowing ample diffusion and time for crystal building, while crystalline ceramics form during high-temperature conditions as dictated by equilibrium phase diagrams.

Geopolymers were first discovered in the Ukraine[1] in the 1950s, where they were referred to as "soil cements." Subsequently they were widely studied for over 30 years by J. Davidovits et al.[2–4] in France, and more recently by several other researchers, notably Van Deventer et al.,[5–7] Barbosa and MacKenzie,[8–11] Balaguru et al.,[12–14] Gauckler et al.,[15–17] Hos et al.,[18] Rahier et al.,[19–21] Comrie et al.[22,23] and Kriven et al.[24] Recent work by Comrie and Kriven[23] demonstrated the potential of reinforced geopolymer composites as thermal-shock-resistant materials.

The aim of this work is to briefly introduce geopolymers and some geopolymer composites and suggest them as possible refractories for versatile applications in the glass manufacturing industry. We describe typical processing procedures; investigate some of their characteristic microstructural features as seen by XRD, SEM, and TEM; and prepare some structural geopolymer composites, reinforced with stainless steel powder, basalt chopped fibers, and/or fiber weaves. The mechanical properties (compressive strength and work of fracture) are to be measured and compared with work of fracture of some common materials.

Table I. Molar oxide ratios for standard geopolymer composition[24]

Components	Molar ratio
SiO_2/Al_2O_3	4:1
K_2O/SiO_2	0.3
H_2O/K_2O	11:1

Experimental Procedures

Processing

A typical, standard method widely used in the literature for preparing a geopolymer paste is based on the idea of dissolving as much silicate as possible in the caustic alikaline solution (NaOH or KOH usually), forming "waterglass." During curing, it is beneficial to seal the geopolymer in plastic wrap to slow down water evaporation from the surface, and hence eliminate surface cracking. The method is as follows:

1. Mix the concentrated alkaline solution NaOH or KOH) with required water and amorphous silica, according to Table I.
2. Allow the silica to dissolve (solution becomes transparent in about 30 min with KOH).
3. Add metakaolin (700°C calcined kaolin) and the remnant silica and any filler phase to the mixture, and stir until homogeneous. After a short time, the mixture thickens.
4. Transfer the mixture to a mold to cure. For uniform microstructure, seal mold in a plastic wrap and place at 40–60°C overnight.

In this study, two clays were used to make geopolymer compression samples after being calcined at 700°C for 1 h. The details of the materials used are given in Table II. In order to make the alkaline solution, KOH pellets from Fisher were dissolved in deionized water. A high-pressure autoclave was set up to produce geopolymer samples that were devoid of macroscopic porosity. In this procedure, the geopolymer suspension was poured into a cylindrical mold and isostatically pressed at 20 MPa while being heated at 80°C for 24 h.

Table II. Materials used and their characteristics

Material (company)	r_{avg}[†] (μm)	SSA[‡] (m^2/g)	Purity	Additional info
Kaolex BN (Kentucky Tennessee Clay Company)	–	22–26	65% kaolin, 7% muscovite, 2.6% TiO_2, 10–30% crystalline SiO_2	71.5% < 0.5 μm, 98% < 5 μm
Hydrite PXN (Dry Branch Kaolin Company)	1.7	15.8	~98% kaolin; impurities: 0.6% Fe_2O_3, 1.4% TiO_2	
Fumed silica (Cabot)	0.2–0.3	380±30	99.8%	Type EH-5
Stainless steel type 316-L (Fisher Scientific)	44	–	99% metals basis, (Fe:Cr:Ni:Mo = 67.5:17:13:2.5)	Stock # 88390, lot # I13M25
KOH pellets (Fisher Scientific)	–	–	89.3% KOH, water as main impurity	P250-3, lot # 030081

[†] r_{avg} = average particle size
[‡] SSA = specific surface area

Table III. Properties of basalt fibers

Physical properties	
Color	Dark brown
Density	2.7 g/cm^3
Hardness	6.5 Mohs
Akaline resistant	Yes
Moisture regain (ASTM D1909)	0
Mechanical properties	
Tensile strength (ASTM D2343)	4840 MPa
Elastic modulus (ASTM D2101)	89 GPa
Elongation at break	3.15%
Thermal properties	
Anneal point (ASTM C338)	648°C
Strain point (ASTM C338)	612°C
Expansion coefficient (D696)	54.7
Melting temperature	1350°C
Operating temperature	Up to 1000°C
Composition (%)	
SiO_2	58.7
Al_2O_3	17.2
CaO	8.04
MgO	3.82
Fe_2O_3	9.72
K_2O	0.82
TiO_2	1.16

Basalt product information from Albarrie Canada Ltd., Process Engineering Division, 85 Morrow Road, Barrie, Ontario L4N 3V7. Website: <www.albarie.com>.

For the composites, the stainless steel powder was also obtained from Fisher. Basalt fibers, in chopped (~10 μm diameter by 4 mm in length) or weave forms, were obtained from Albarrie Canada Ltd., Barrie, Ontario, Canada. Basalt is a dense, dark volcanic rock that has good thermal properties, strength, and durability. The basalt has been formed into continuous fibers via a single stage melt process. The fibers were coated with silane sizing, which was left on during processing and in the final composite. Table III lists some physical characteristics and the typical chemical compositions of basalt fibers.

Microstructure Characterization

Samples of geopolymer were characterized using XRD of powders made from crushed geopolymer bodies. SEM specimens were prepared by sputter coating the surface with a gold palladium coating. The specimen preparation for TEM deviated from the normal procedure because of the emission of water even at low temperatures achieved on a hot plate (e.g., 120°C). Such heat treatments caused the pure geopolymer to become brittle and difficult to handle. Therefore, heat treatment of the geopolymer was avoided by using Krazy Glue® and ion milling under liquid nitrogen conditions in a Gatan ion miller.

Scanning electron microscopy (SEM) was performed on as-prepared and fractured samples using a Zeiss DSM-960.* SEM studies were performed on a Hitachi S-800 field emission SEM.

Transmission electron microscopy (TEM/EDS) studies were conducted on a Philips CM 12 instrument† equipped with energy dispersive X-ray spectroscopy (EDS) and operated at 100 KV. An ion-milled, 3 mm disk of pristine, unheated geopolymer was examined.

Mechanical Evaluation

Geopolymer samples were prepared either as a single phase or reinforced with stainless steel by mixing the metal into the waterglass (alkali-silicate solutions) with metakaolin. Extra stirring was needed in order to homogenize the metal particles throughout the resulting geopolymer paste. The mixture was then cast into selected molds, covered, and cured at 45°C for 24 h. Compression testing cylinders (0.25 in. diameter by 1 in. length) and rectangular bend bars (1 × 1 × 6 in. length) were made and tested. A number of control samples were tested for comparison purposes. Compression samples and four-point bend testing was done in accordance with ASTM C 773-88 and ASTM C 1161-94. The machine used to test the compression cylinders was an Instron 4502.‡ The bend bars were tested using a Tinius Olsen machine under four-point bending.

Natural basalt fibers fibers were added to the geopolymer samples according to Table IV. Three types of samples were made: compression cylinders (0.25 in. diameter by 1 in. length), rectangular bend bars (1 × 1 × 6 in. length), and plate pressed composites. Only chopped fibers were

*Thornwood, New York.

†Philips Inc., Mahwah, New Jersey.

‡Canton, Massachusetts.

Table IV. Fibers added during processing

Material (company)	Dimensions	Samples made
Basalt chopped fiber (Kentucky Tennessee Clay Company)	8–11 μm (D) × 2 mm (L)	Bend bars, cylinders
Basalt fiber mesh (Dry Branch Kaolin Company)	0.75 mm thick — used 1 × 1 × 7 in. layers	Bend bars, plates
Basalt fiber and mesh	–	Bend bars
Carbon fiber mesh (McMasterCarr)	1-in. width	Plates

added to the compression cylinder samples, because of their small size. The cylinders were tested in compression in a similar manner as the metal reinforced samples. An Instron 8800* was used to test the bend bars with the ability to plot load against displacement. The plate pressed samples were not tested mechanically. Kaolex BN and Cabosil silica were used to make all samples. In all cases the molar oxide ratios of materials used were $SiO_2:Al_2O_3 = 4$, $K_2O:SiO_2 = 0.3$, and $H_2O:K_2O = 11$.

Figure 2. Geopolymers can be formed in plastic molds and have a smooth surface finish.

Results

It is known that geopolymers do not bond to plastics or graphite, which opens the way to versatility in processing of complex shapes. Figure 2 demonstrates that geopolymers can be formed in plastic molds and have a smooth surface finish. It has also been demonstrated that geopolymers can be reinforced with graphite fibers or fiber weaves, although carbon itself does not bond to geopolymers (Fig. 3).

*25/50 KIP load frame, model 311.15, serial 159.

Microstructure Characterization

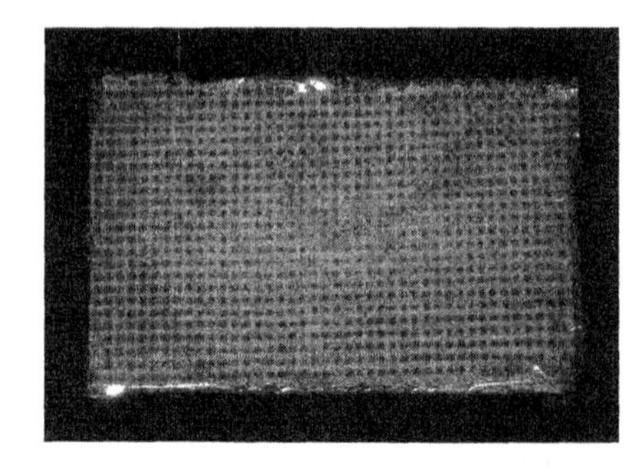

Figure 3. Geopolymers can be reinforced with carbon fiber and fiber weaves.

The XRD diffractometry plots of metakaolinite and converted potassium-cross-linked geopolymer are seen in Fig. 4. Both phases are amorphous, with a characteristic hump centered about ~22° in 2θ in the metakaolin. After geopolymerization, the amorphous hump is moved to ~28° in 2θ. This latter feature seems to be characteristic of reacted geopolymer, irrespective of the alkali metal (Na, K, Cs) used in the composition.

In SEM micrographs of geopolymer made from natural metakaolin, there is always evidence of unreacted phase, as seen from aligned plates of metakaolin (Fig. 5). However, in higher resolution SEM, the microstructure of the geopolymer in the fully reacted material is very fine, homogeneous, and nanoparticulate (Fig. 6). Corresponding TEM brightfield micrographs

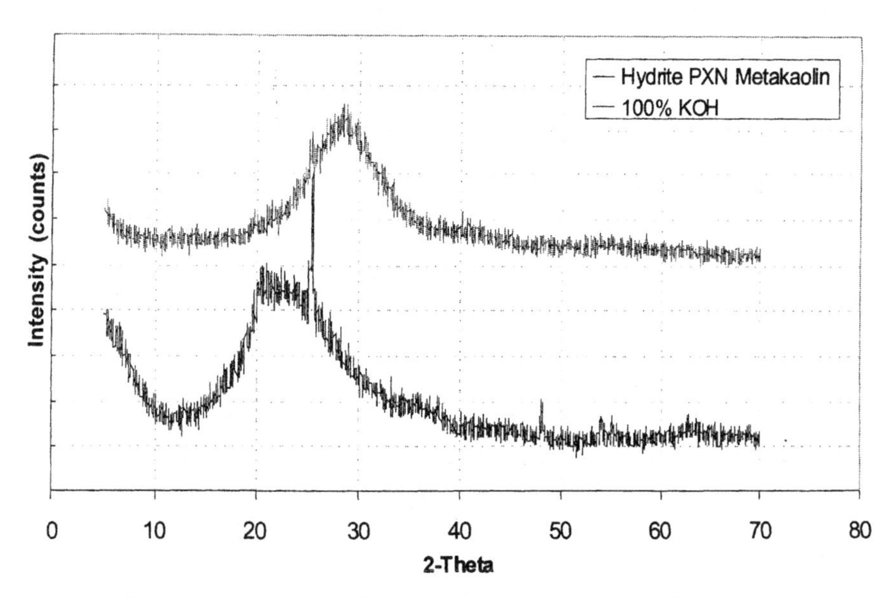

Figure 4. XRD plot of hydrite PXN metakaolin before (lower plot) and after (upper plot) conversion to geopolymer. The dehydrated metakaolin has a charcteristic amorphous hump at ~22° 2θ and shifts to ~28° 2θ, which is a reproducible feature of geopolymerization.

Figure 5. SEM micrograph of Na-cured geopolymer made from metakaolinite. Average particle size was 2 µm diameter and specific surface area was 12 m^2/g.

Figure 6. Higher magnification SEM micrograph of fully reacted region of Na-cross-linked polysialate. The starting metakaolinite average particle size was 2 µm diameter and specific surface area was 12 m^2/g.

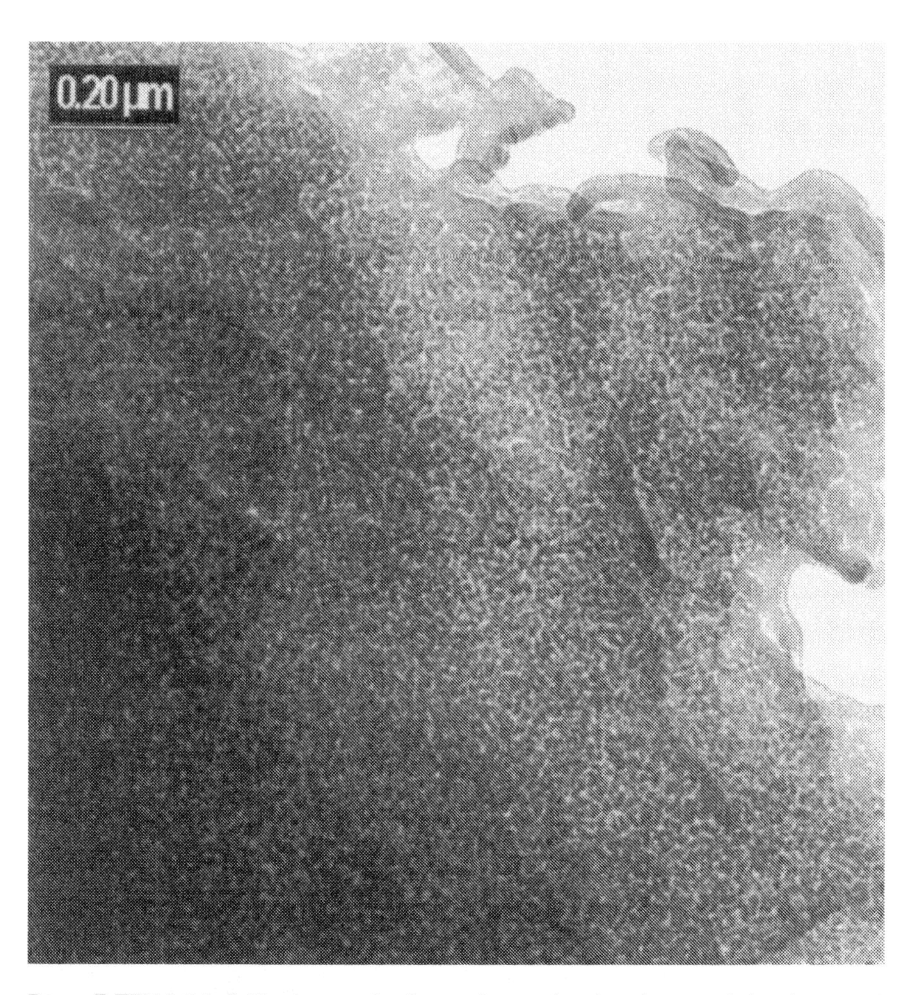

Figure 7. TEM brightfield micrograph of geopolymer showing characteristic microstructure of nanoprecipitated, nanoporous, K-cross-linked polysialate.

confirm the amorphous nature of the microstructure as seen by conventional elected area diffraction patterns. However, the microstructure has a characteristic nanoparticulate, nanoporous texture as seen in Figs. 7 and 8.

Mechanical Properties of Unreinforced Geopolymer

In order to study the effect of alkaline choice, six bend bar samples for compression were made from each of the following three hydroxides: NaOH, KOH, and CsOH. Samples were prepared using pressureless curing. Compression testing results are shown in Fig. 9. The KOH had the best

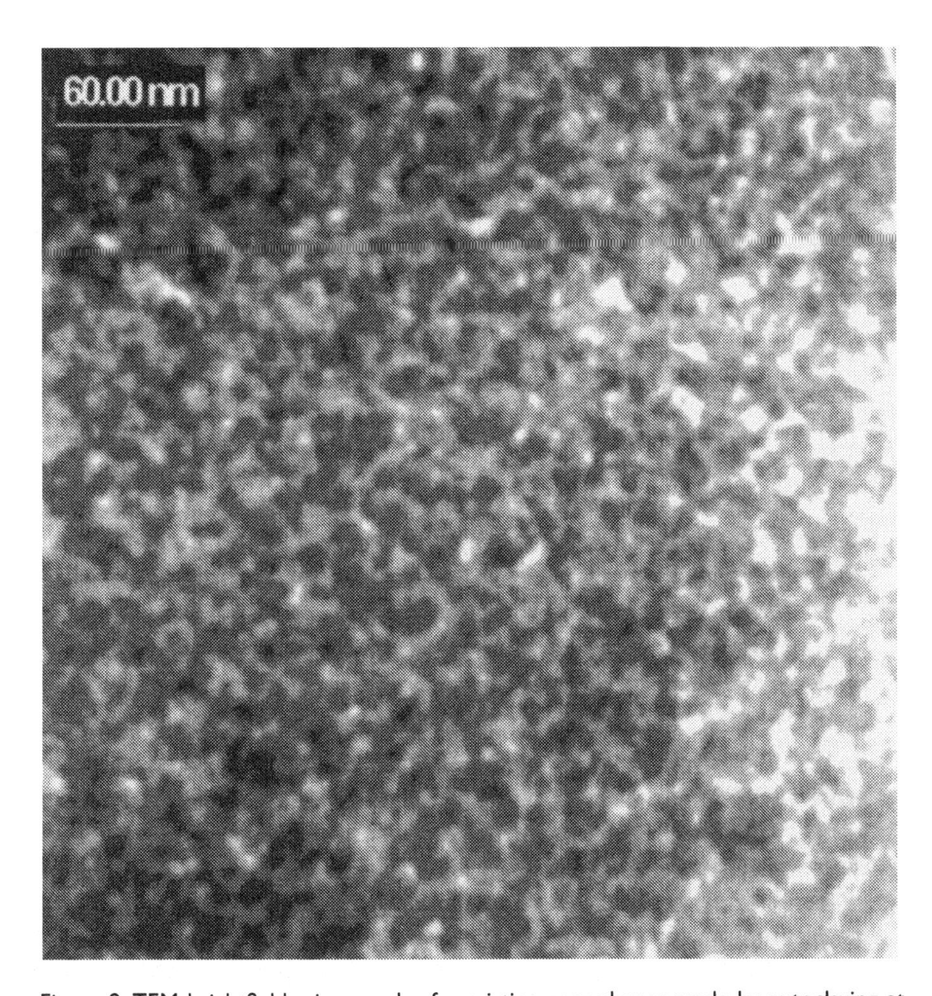

Figure 8. TEM brightfield micrograph of a pristine geopolymer made by autoclaving at 80°C a metakaolin powder with an average particle size and surface area of 0.3 μm and 18 m^2/g, respectively, which shows nearly complete reaction. This material corresponds to the SEM micrograph of Fig. 5. The microstructure appears to be lumpy and bumpy, with reacted globules of diameter on the order of less than 10 nm. Examination in a higher resolution microscope may shed more light on the polymerized microstructure of a single globule. The average composition as measured by TEM/EDS is $4SiO_2{\cdot}Al_2O_3{\cdot}K_2O$.

compressive strength, followed by NaOH and CsOH, respectively. SEM analysis of geopolymers made from different alkali hydroxides confirmed that cation size influenced the dissolution of aluminosilicate starting materials as shown in Fig. 10. Smaller cations led to greater dissolution, that is,

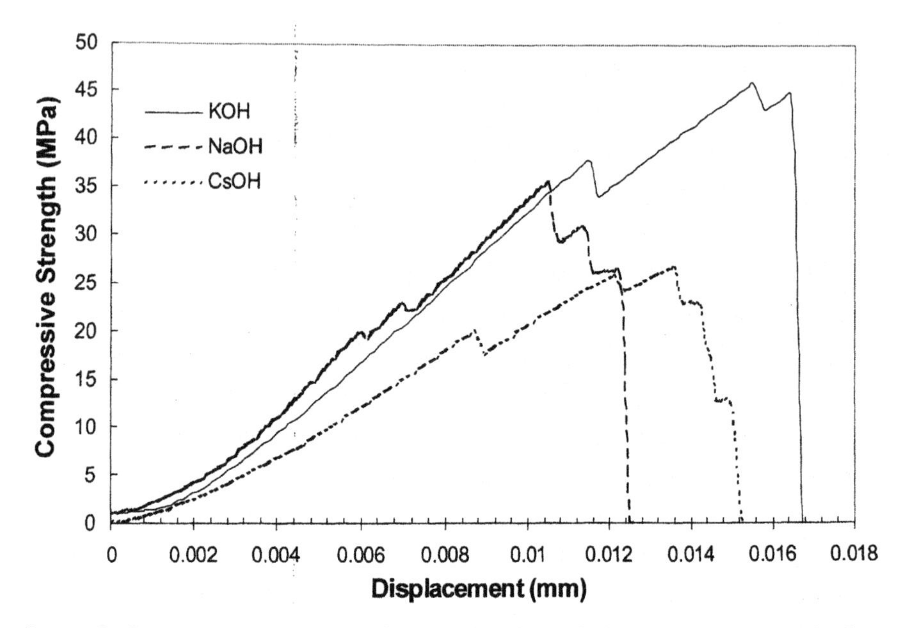

Figure 9. Compression testing results as a function of alkaline choice. A total of six samples were made using each alkaline choice. CsOH was plotted twice.

cation size Na < K < Cs. More in depth work has recently been done by Kriven and Bell[25] on the effects of Na versus K versus mixtures of both on the porosity and mechanical properties of the resulting geopolymers.

The results from compression testing of cylinders were plotted as load versus displacement. Control samples without any filler had strength values ranging from 40 to 60 MPa in 80°C autoclaved compression cylinders. By changing the chemistry and pulling a vacuum to remove entrapped air, values as high as 83 MPa with an average value of 58 MPa could be reached for 14 tested samples. However, for this study, a vacuum was not applied to any of the samples. The clay used was a factor in the resulting compressive strength of the samples. Samples made from Kaolex BN were not as fully reacted as compared to those made from purer kaolin clays. The Kaolex BN contained 7% muscovite and up to 30% crystalline silica. These phases may have been left unreacted in the final material, thus acting as filler. Additional testing was done using metakaolin derived from 98% pure kaolin clay (Hydrite PXN*). Strength values were in the 20–50 MPa range using Hydrite PXN when the mix was exposed to a vacuum. The samples

*Dry Branch Kaolin Company.

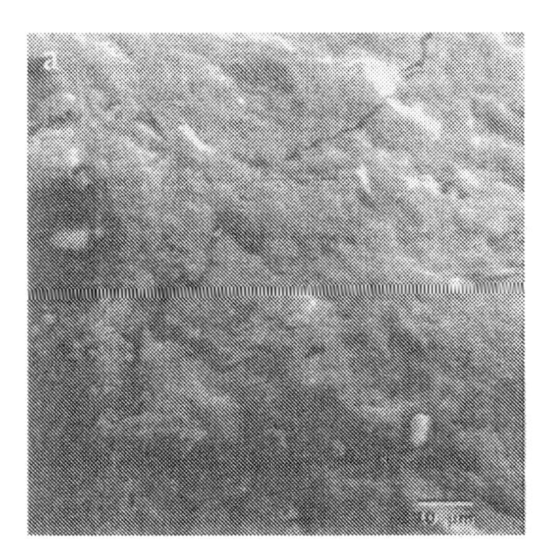

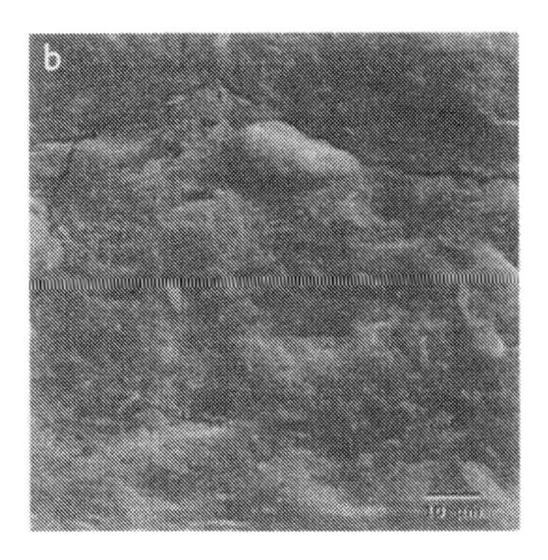

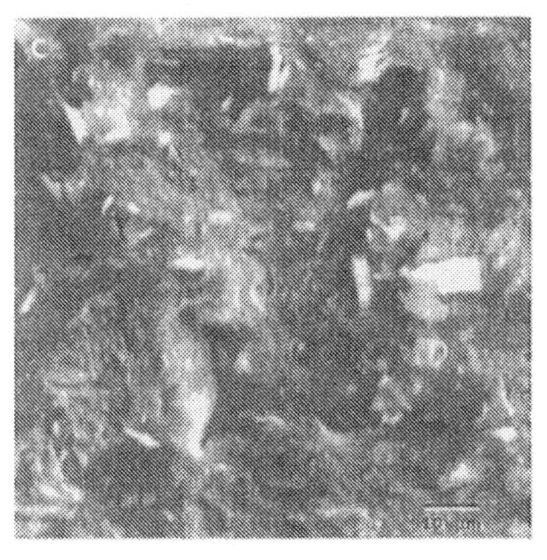

Figure 10. SEM pictures of fracture surfaces of tested compression cylinders: (a) NaOH, (b) KOH, (c) CsOH. A significant amount of undissolved material was left behind when using CsOH.

had an average strength of 36.5 MPa for 13 samples. A summary of these numbers is given in Table V. Work of fracture values for some other common materials are listed in Table VI for comparison purposes.

Results for bend bar testing were plotted as bending strength (MPa) vs. displacement (mm) for all samples. Six samples were tested for each sample type. The control samples with no filler phase fractured cleanly and abruptly at stresses less than 3 MPa. The work of fracture ranged from 0.01 to 0.07 kJ/m^2. An optical micrograph of a bend bar containing no added filler phase is shown in Fig. 11, wherein it is seen that the samples fractured cleanly and abruptly, as is typical of a ceramic.

Table V. Compressive strength results for control samples

Clay used	Vacuum pulled	Number of samples tested	Compressive strength range (MPa)	Average compressive strength (MPa)
Kaolex BN	No	12	40–58	45
Kaolex BN	Yes	14	40–83	59
Hydrite PXN	Yes	13	21–48	37

Table VI. Work of fracture values for common materials[26]

Material	Work of fracture (kJ/m^2)
Copper	50
Key steel	50
Brass	30
Teak wood	6
Cast iron	4
Toughened polystyrene	4
Deal wood	2
Cellulose	2
Polystyrene	1
Reactor graphite	0.1
Firebrick	0.02–0.07
Alumina	0.04

Mechanical Properties of Reinforced Geopolymer Composites

Stainless steel addition led to enhanced strength and toughness (Fig. 12). The maximum ultimate compressive strength of 76 MPa and bend strength of 28 MPa were achieved at 15 vol% addition. Above 15 vol% addition, strength and toughness values began to decrease. It is thought that the stainless steel increased the strength by blunting propagating cracks. Unreacted metakaolin regions were also observed in the stainless steel–filled samples.

Chopped basalt fibers (4 mm in length) were added to compression cylinders at 0, 1, and 5 vol%. The pourability of the paste was lost beyond 1 vol% fiber addition and it had to be forced in to the mold. By adding the fiber, the fracture behavior of the samples changed (Fig. 13). Parts of the specimens chipped off in a pseudoductile manner, which became more pro-

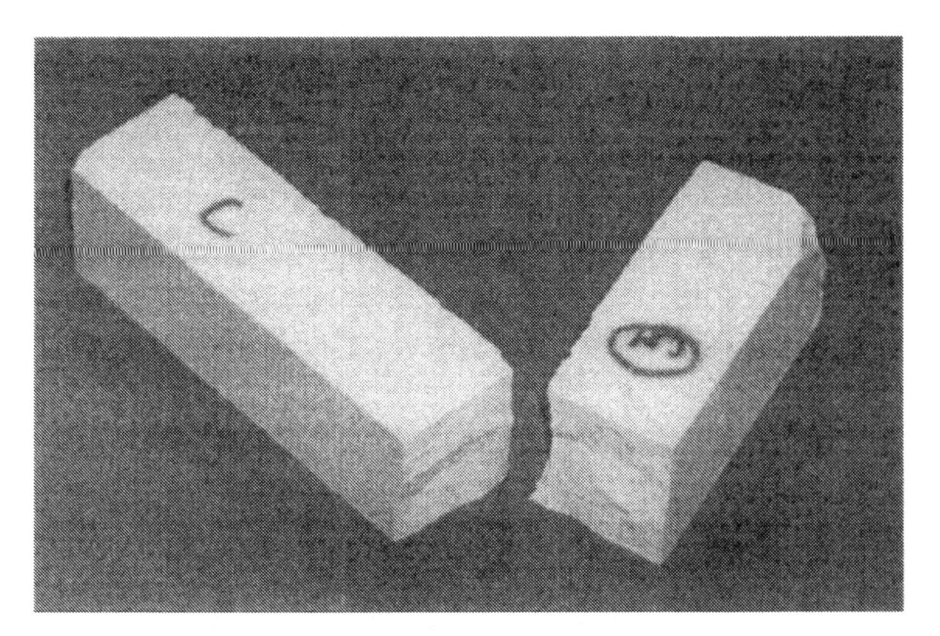

Figure 11. Optical micrograph of a 1 × 1 × 6 in. bend bar of almost fully reacted geopolymer phase made from metakaolin cross-linked from KOH solution. The bend bar exhibited ceramiclike brittle fracture.

nounced with increasing chopped fiber additions. The ultimate strength was reduced, and the head displacement prior to final fracture was drastically increased.

The addition of basalt fiber weave increased both ultimate stress and work of fracture values. The results for samples reinforced with two layers of weave are shown in Fig. 14. The ultimate strength more than tripled compared to the control samples, and the work of fracture (WOF) nearly reached 20 KJ/m^2. Some of the tests did not run to completion because the bars bent beyond the range of the testing fixture. None of the samples fractured cleanly and each had significant fiber pullout (Fig. 15). There were cracks throughout the samples with areas chipped off in some regions. Bend bars were also made using six layers of basalt mesh (Fig. 16). This led to the maximum increase in WOF observed in this study of ~22kJ/m^2 and a slight increase in ultimate stress. It would have been difficult to add more than six layers of mesh without using pressure.

The addition of basalt chopped fibers to the two-weave layer reinforced composite also increased the ultimate stress (~12 MPa) and WOF (4.93 ±

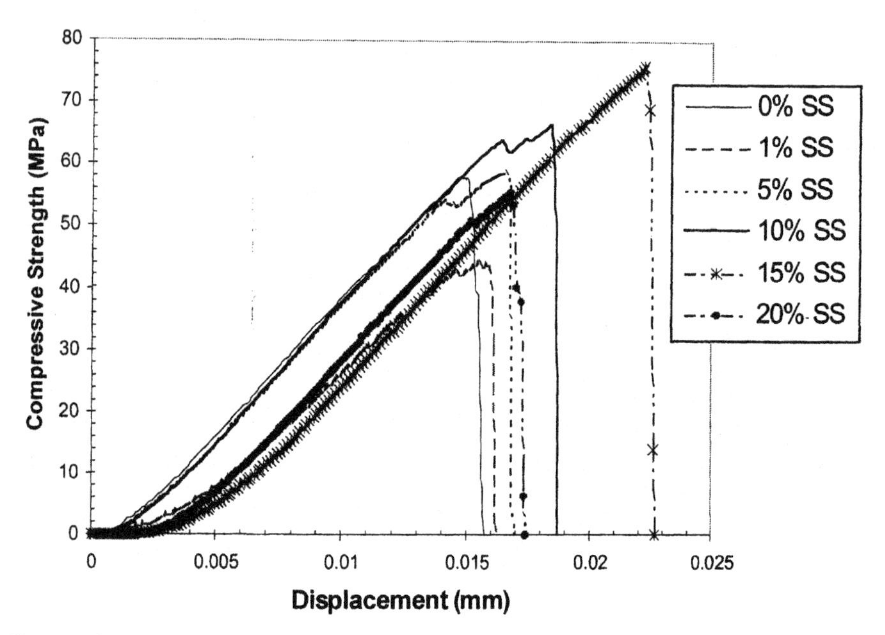

Figure 12. Load vs. displacement for stainless steel–reinforced geopolymer cylinders. Optimum compressive strength was attained at 15 vol% addition of stainless steel particles.

1.50 MPa), as seen in Fig. 17. However, the WOF was much lower than that of the fiber mesh samples. Two of the six tested samples fractured at very low stress. Overall, the data was more sporadic with the chopped fibers compared to the weave or the unreinforced control samples. This may have been due to the poor workability of the paste leading to significant defects in the final samples.

Samples made using both basalt chopped fiber at 1.7 vol% and two layers of basalt fiber weave had a high ultimate strength but a lower WOF compared to samples reinforced with only the weave. WOF values were similar to bend bars reinforced with only chopped fiber. Chopped fibers caused the paste to become unpourable, leading to poor sample workability. It is also likely that the weave was not well bonded to the geopolymer matrix.

Plate pressed, multilayer, fiber weave composites were made using a warm press. A carbon fiber composite made using the plate press is shown in Fig. 18.

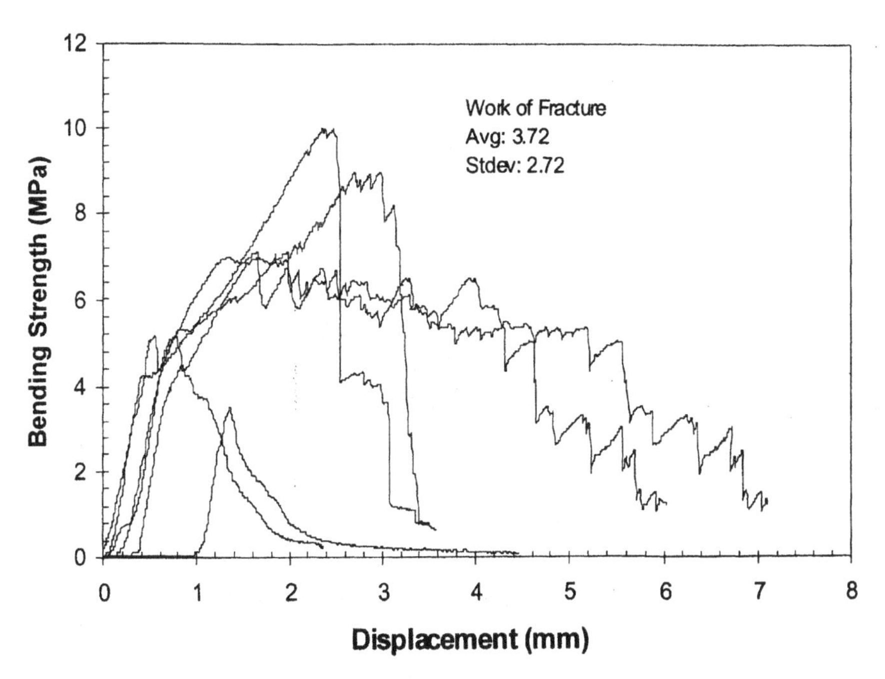

Figure 13. Stress vs. displacement curves for bend bars reinforced with 1.7 vol% of basalt chopped fiber.

Discussion and Conclusion

This work has briefly introduced geopolymers and some geopolymer composites from the perspective of potential applications as refractories applications in the glass manufacturing industry. Typical compositions in terms of molar oxide ratios of materials used were SiO_2:Al_2O_3 = 4, M_2O:SiO_2 = 0.3, and H_2O:M_2O = 11, where M = Na, K, or Cs. Using standard processing procedures, 1 × 2 × 6 in. long bend bars and 0.25 × 1 in. long cylinders for compression testing were fabricated. Some of their characteristic microstructural features as seen by XRD, SEM, and TEM were examined. The microstructure consists of a nanoparticulate (< 10 nm) and nanoporous material having a nominal composition of $4SiO_2 \cdot Al_2O_3$ cross-linked with cations from Na_2O, K_2O, or Cs_2O. Both single phase as well as geopolymer composites were made, for example, those reinforced with stainless steel powder, basalt chopped fibers, and/or fiber weaves. The mechanical properties (compressive strength and work of fracture) were measured and compared with work of fracture of some common materials. The mechanical

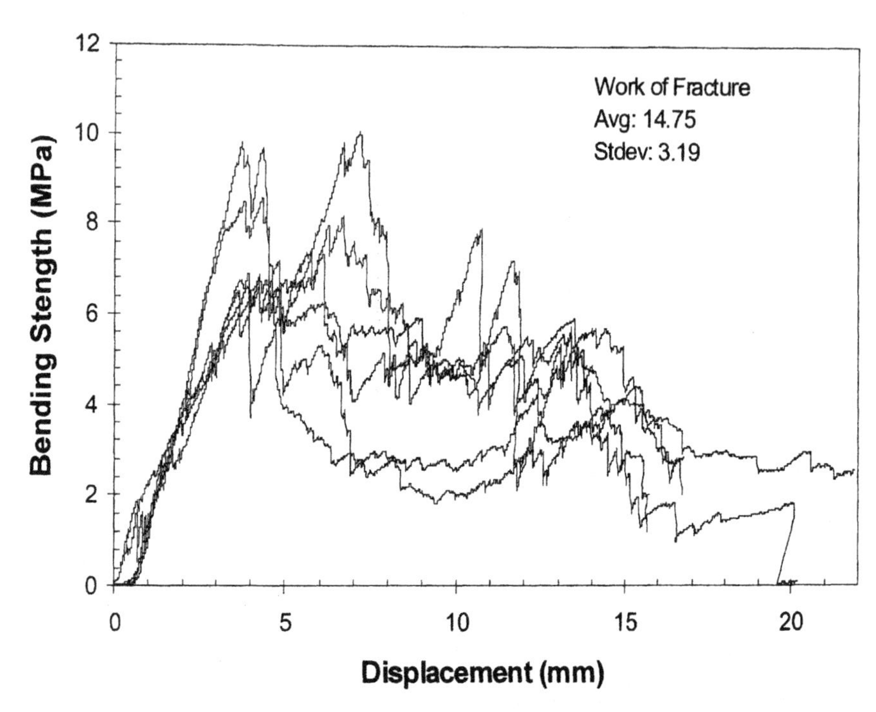

Figure 14. Fracture behavior of bend bars reinforced with two layers of basalt mesh. The samples were cracked in multiple regions and took a long time to test.

strength of unreinforced geopolymer compression cylinders varied between 40 and 60 MPa in samples autoclaved at 80°C under 20 MPa at 80°C for 24 h. However, bend bars made by pressureless curing at ambient pressures had low intrinsic bend strengths of less than 3 MPa, with an average work of fracture of 0.05 kJ/m^2. Geopolymers reinforced with 15 vol% stainless steel powder had a maximum compressive strength of ~76 MPa and bend strength of ~28 MPa. Bend bars reinforced with 1.7 vol% basalt chopped fiber broke after an average work of fracture of 3.7 ± 2.7 KJ/m^2. Six-inch bend bars reinforced with two layers of basalt weave had a maximum strength of ~10 MPa and work of fracture of 14.75 ± 3.19 kJ/m^2. Using six layers of sized natural basalt fiber weave as reinforcements, the three-point bend strength and work of fracture of 1 × 2 × 6 in. long bend bars increased from an average of 2.75 to 10.25 MPa and from 0.05 to 21.82 kJ/m^2, respectively. Thus, significant strength and toughness enhancements can be achieved by reinforcing geopolymers made under ambient temperature and pressure conditions.

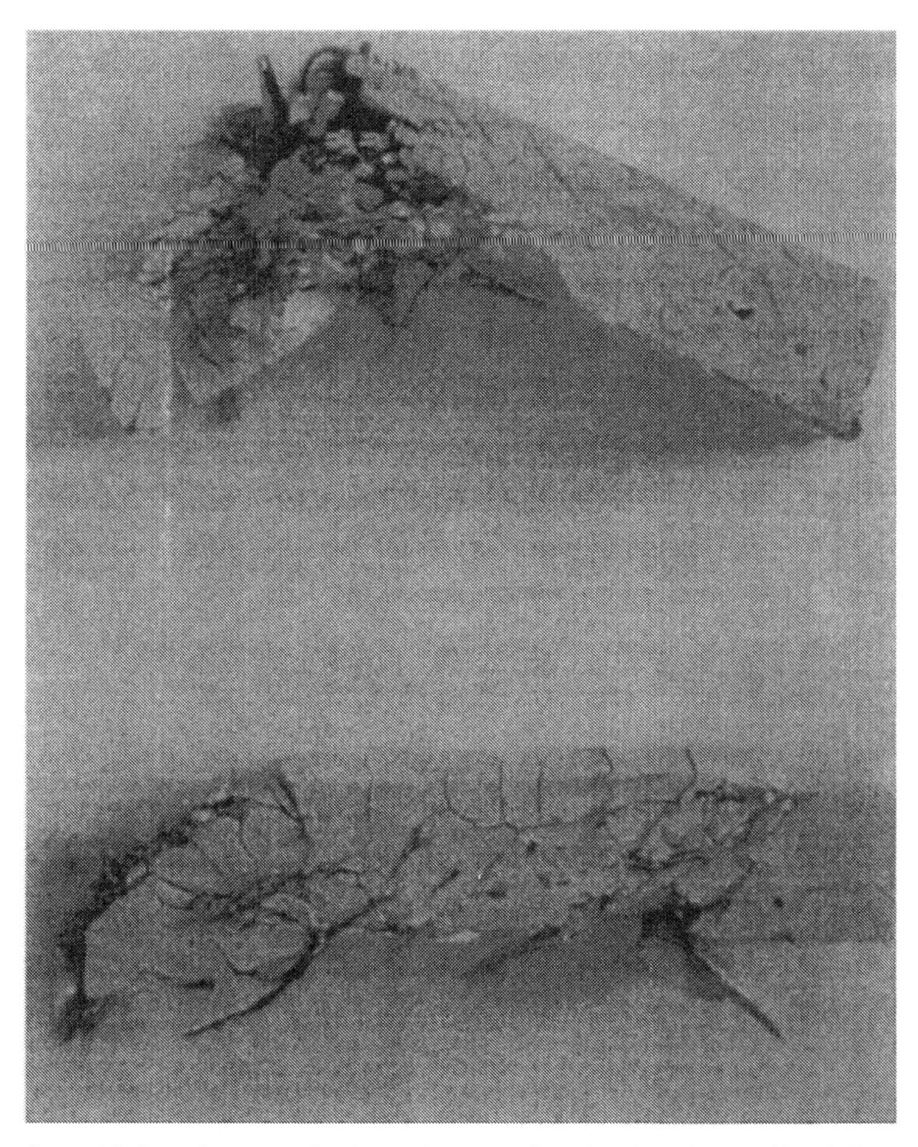

Figure 15. Optical micrograph of geopolymer reinforced with six layers of basalt fiber weave. The maximum work of fracture was measured at ~22 kJ/m^2.

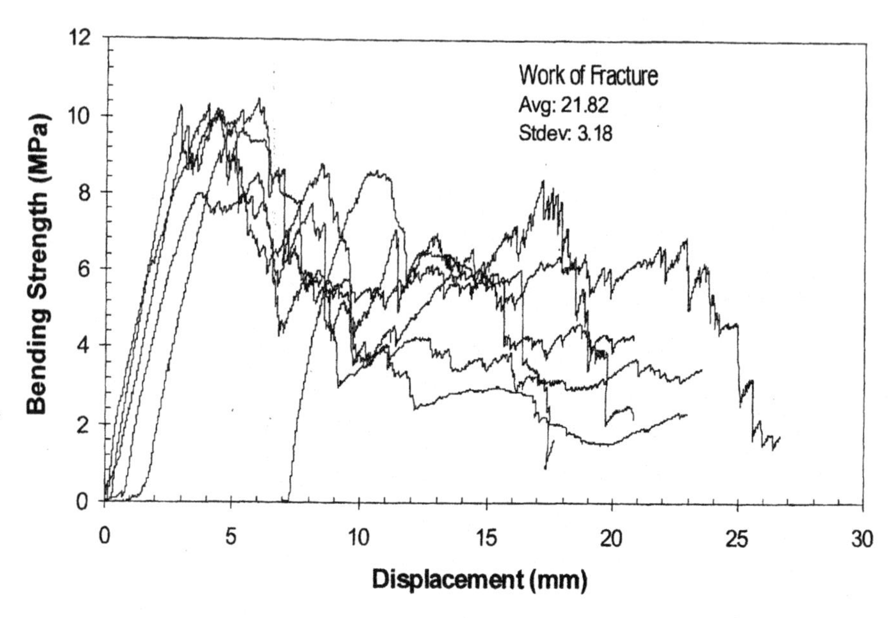

Figure 16. Stress vs. displacement curves for bend bars reinforced with six layers of basalt mesh. A maximum work of fracture of 21.82 kJ/m^2 was realized in this study.

Acknowledgments

This work was supported by the AFOSR, under STTR grant number F49620-02 C-010. The authors acknowledge the invaluable assistance of Dr. G. Lukey and Prof. J. Van Deventer from the University of Melbourne, Australia, in identifying some of the key background literature references. The authors also thank Dr. J. Davidovits of St. Quentin, France; Dr. K. J. D. Mackenzie from the New Zealand Institute for Industrial Research and Development, New Zealand; as well as Mr. D. C. Comrie from Catawba Resources, Inc., Ohio, for insightful discussions.

References

1. V. D. Glukhovsky, "Ancient, Modern and Future Concretes"; pp. 1–9 in *Alkaline Cements and Concrete,* Vol. 1. Edited by P. V. Krivenko. VIPOL Stock Company, Kiev, Ukraine, 1994.
2. J. Davidovits, "Geopolymers and Geopolymeric Materials," *J.Thermal Anal.,* **35** [2] 429–441 (1989).
3. J. Davidovits, "Geopolymers: Inorganic Polymeric New Materials," *J. Thermal Anal.,* **37** [8] 1633–1656 (1991).

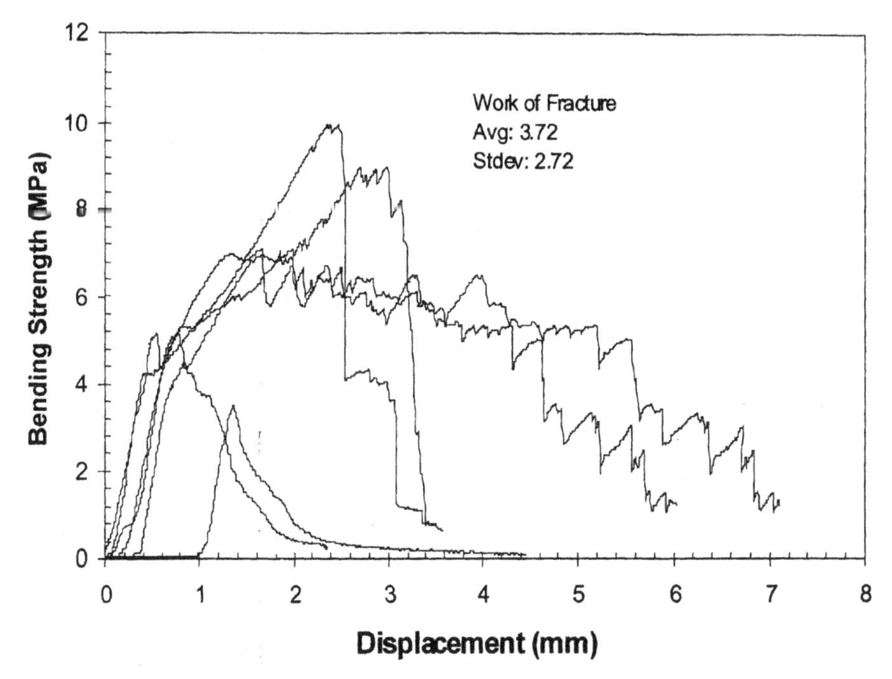

Figure 17. Stress vs. displacement curves for bend bars reinforced with 1.7 vol% of basalt chopped fiber, having an average work of fracture of 3.72 kJ/m^2.

4. J. Davidovits, "Geopolymers, Man-Made Rock Geosynthesis and the Resulting Development of Very Early High Strength Cement," *J. Mater. Educ.*, **16** [2–3] 91–137 (1994).
5. H. Xu and J. S. J. van Deventer, "The Geopolymerisation of Alumino-Silicate Minerals," *Int. J. Mineralogical Processes,* **59** [3] 247–266 (2000).
6. H. Xu, J. S. J. Van Deventer, and G. C. Lukey, "The Effect of Alkali Metals on the Preferential Geopolymerisation of Stilbite/Kaolinite Mixtures," *Indian Eng. Chem. Res.*, **40** [17] 3749–3756 (2001).
7. C. K. Yip, G. C. Lukey, and J. S. J. van Deventer, "Effect of Blast Furnace Slag Addition on Microstructure and Properties of Metakaolinite Geopolymeric Materials"; pp. 187–209 in *Advances in Ceramic Matrix Composites IX.* Ceramic Transactions vol. 153. Edited by N. P. Bansal, J. P. Singh, W. M. Kriven, and H. Schneider. American Ceramic Society, Westerville, Ohio, 2003.
8. V. F. F. Barbosa, K. J. D. MacKenzie, and C. Thaumaturgo, "Synthesis and Characterisation of Materials Based on Inorganic Polymers of Alumina and Silica: Sodium Polysialate Polymers," *Int. J. Inorganic Materials,* **2**, 309–317 (2000).
9. V. F. F. Barbosa and K. J. D. MacKenzie, "Thermal Behavior of Inorganic Geopolymers and Composites Derived from Sodium Polysialate," *Mater. Res. Bull.*, **38**, 319–331 (2003).

Figure 18. Multilayer basalt fiber mesh composite made by warm pressing eight layers of basalt fiber. The picture on the left presents a side view; the picture on the right is a top view.

10. V. F. F. Barbosa and K. J. D. MacKenzie, "Synthesis and Thermal Behavior of Potassium Sialate Geopolymers," *Mater. Lett.*, **57**, 1477–1482 (2003).
11. K. J. D. MacKenzie, "What Are These Things Called Geopolymers? A Physico-Chemical Perspective"; pp. 175–186 in *Advances in Ceramic Matrix Composites IX.* Ceramic Transactions vol. 153. Edited by N. P. Bansal, J. P. Singh, W. M. Kriven, and H. Schneider. American Ceramic Society, Westerville, Ohio, 2003.
12. R. E. Lyon, U. Sorathia, P. N. Balaguru, A. Foden, J. Davidovits, and M. Davidovits, "Fire Response of Geopolymer Structural Composites"; pp. 972–981 in *Proceedings of 1st International Conference on Fiber Composites Infrastructure.* 1996.
13. R. E. Lyon, P. N. Balaguru, A. Foden, U. Sorathia, J. Davidovits, and M. Davidovits, "Fire-resistant Aluminosilicate Composites," *Fire Mater.*, **21** [2] 67–73 (1997).
14. C. G. Papakonstantinou, P. N. Balaguru, and R. E. Lyon, "Comparative Study of High Temperature Composites," *Composites: Part B*, **32**, 637–649 (2001).
15. S. C. Förster, T. J. Graule and L. J. Gauckler, "Strength and Toughness of Reinforced Chemically Bonded Ceramics"; pp. 247–256 in *Cement Technology.* Ceramic Transactions vol. 40. Edited by E. M. Gartner and H. Uchikawa. American Ceramic Society, Westerville, Ohio, 1994.

16. S. C. Förster T. J. Graule, and L. J. Gauckler, "Thermal and Mechanical Properties of Alkali-Activated Alumino-Silicate Based, High-Performance Composites"; pp. 117–124 in *Advanced Structural Fiber Composites.* Advances in Science and Technology vol. 7. Edited by P. Vincenzini. Techna Srl., 1995.
17. S. C. Förster, "Alkaliaktivierte Aluminosilikat-verbundkeramiken" [Alkali-Activated, Aluminosilicate-Bound Ceramics] (in German), Diplom Engineur Thesis, Dissertation ETH No.10773, Zurich, Switzerland, 1994.
18. J. P. Hos, P. G. McCormick, and L. T. Byrne, "Investigation of a Synthetic Aluminosilicate Inorganic Polymer," J. Mater. Sci., **37**, 2311–2316 (2002).
19. H. Rahier, B. Van Mele, M. Biesemans, J. Wastiels, and X. Wu, "Low-Temperature Synthesized Aluminosilicate Glasses. Part I. Low-Temperature Reaction Stoichiometry and Structure of a Model Compound," *J. Mater. Sci.,* **31**, 71–79 (1996).
20. H. Rahier, B. Van Mele, and J. Wastiels, "Low-Temperature Synthesized Aluminosilicate Glasses. Part II. Rheological Transformations During Low-Temperature Glasses," *J. Mater. Sci.,* **31**, 80–85 (1996).
21. H. Rahier, W. Simons, B. Van Mele, and M. Biesemans, "Low-Temperature Synthesized Aluminosilicate Glasses. Part III," *J. Mater. Sci.,* **32**, 2237–2247 (1997).
22. D. C. Comrie and J. Davidovits, "Long Term Durability of Hazardous Toxic and Nuclear Waste Disposals"; pp. 125–134 in *Geopolymer '88,* vol. 1. Proceedings of the First European Conference on Soft Mineralurgy, Compiegne, France, 1988.
23. D. C. Comrie and W. M. Kriven, "Composite Cold Ceramic Geopolymer in a Refractory Application"; pp. 211–255 in *Advances in Ceramic Matrix Composites IX.* Ceramic Transactions vol. 153. Edited by N. P. Bansal, J. P. Singh, W. M. Kriven, and H. Schneider. American Ceramic Society, Westerville, Ohio, 2003
24. W. M. Kriven, J. L. Bell, and M. Gordon, "Microstructure and Microchemistry of Fully Reacted Geopolymers and Geopolymer Matrix Composites"; pp. 227–250 in *Advances in Ceramic Matrix Composites IX.* Ceramic Transactions vol. 153. Edited by N. P. Bansal, J. P. Singh, W. M. Kriven, and H. Schneider. American Ceramic Society, Westerville, Ohio, 2003).
25. W. M. Kriven and J. L. Bell, "Effect of Alkali Choice on Geopolymer Properties," *Ceram. Eng. Sci. Proc.,* **25** (2004).
26. H. G. Tattersall and G. Tappin, "The Work of Fracture and Its Measurement in Metals, Ceramics and Other Materials," J. Mater. Sci., **1**, 296–301 (1966).

Anomalous Thermomechanical Properties of Network Glasses

John Kieffer and Liping Huang*
Department of Materials Science and Engineering, University of Michigan, Ann Arbor, Michigan

Anomalous thermomechanical behaviors of silica glass, which include negative thermal expansion and the increase of elastic moduli with decreasing density, have been known for some time but no satisfactory explanation has been available. Recent advances in measurement technology and computational means for interpreting the observed behaviors have allowed us to improve our understanding of these phenomena. In this paper we present experimental data showing that anomalous temperature dependences of mechanical properties are common to all strong network glass formers, and that they extend into the liquid state. Furthermore, based on atomic-scale computer simulations we provide an explanation for the anomalous properties of silica.

Introduction

It is well known that the properties of strong network glass formers, of which silica can be considered the archetype, exhibit anomalous temperature and pressure dependencies. These are more or less pronounced depending on the thermomechanical conditions, and some of these behaviors even persist in mixtures with modifying compounds for as long as a significant degree of networking remains in the glass structure. For example, at very low temperatures and again at temperatures far above the glass transition silica exhibits negative thermal expansion.[1] Furthermore, the elastic moduli of silica glass increases with increasing temperature,[2–4] and the bulk modulus passes through a minimum upon compression at ~2–3 GPa.[5–10] This material can also undergo irreversible densification under pressure.[11–16]

In the case of silica these anomalous behaviors have been known for several decades, although a generally accepted explanation had not been presented yet. In this paper we report the results of atomic-scale computer simulations that reveal the mechanisms that underlie these behaviors. This study not only provides a resolution for a long-standing puzzle in glass science, but it also asserts the emerging predictive capabilities of atomic-scale

*Currently on leave from the Department of Materials Science and Engineering, University of Illinois.

computer simulations. The paper is organized as follows. First, the experimental observations of anomalous thermomechanical behavior will be briefly reviewed. Then follows a brief introduction in the methodology of molecular dynamics simulations. The research strategy is outlined after that. Finally, simulation results are presented, and based upon these an explanation for the anomalous temperature and pressure dependencies of the mechanical properties of silica is derived.

Measurement of Mechanical Properties at High Temperatures and Pressures

Although silica is perhaps the most prevalent glass former from a technological point of view, it is important to realize that the aforementioned anomalous thermomechanical properties are ubiquitous for all strong network formers. This can be seen in Fig. 1, where the longitudinal and shear elastic moduli of SiO_2, GeO_2, and B_2O_3 are plotted as a function of temperature.[17] All three of these glass formers exhibit a positive temperature dependence of their elastic moduli, at least over a significant portion of the explored temperature range.

The mechanical properties shown in Fig. 1 were measured using the Brillouin light scattering technique. This technique works on the basis that light scatters from condensed phases because of density fluctuations, such as sound waves. Sound waves exist in any material at finite temperature because they are the mechanism by which the system stores thermal energy. When monochromatic light scatters off propagating sound waves, a Doppler shift is introduced in the spectrum of scattered light, which can be resolved using a sensitive interferometer. Typical magnitudes are on the order of 10 GHz. This Doppler shift is proportional to the sound velocity, and consequently, to the square root of the elastic modulus.

In principle, because the mechanical properties are probed by an oscillatory deformation (sound waves in this case), the resulting measure is a complex quantity, which has a real and an imaginary component. The real component is the elastic storage modulus and the imaginary component is the loss modulus. The latter can be seen as the result of sound waves being attenuated as they propagate. For liquids, the loss modulus can be directly related to the viscosity of the scattering medium. Hence, the Brillouin technique provides a means for determining the viscoelastic properties of glasses and glass-forming melts.

There are several important attributes to the way the elastic moduli and

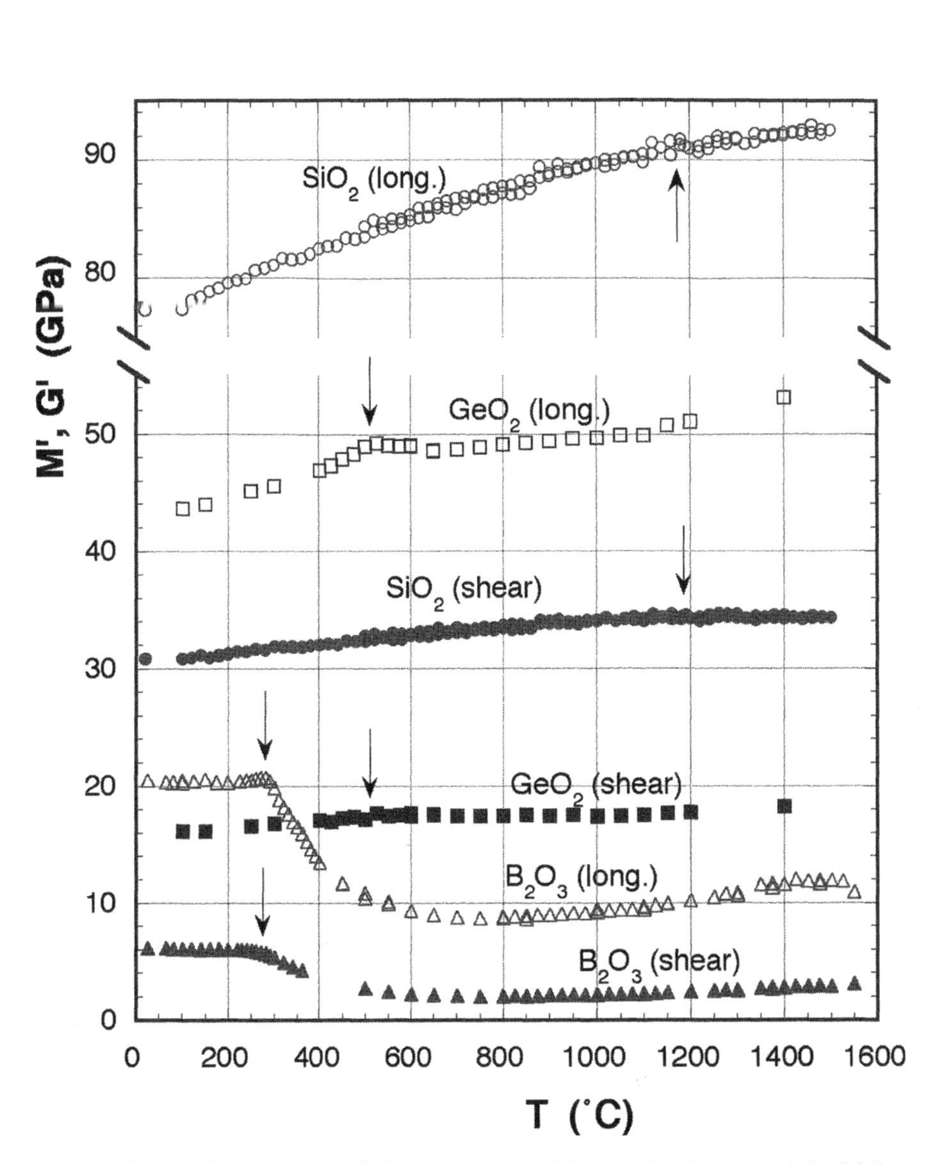

Figure 1. Longitudinal (open symbols) and shear (solid symbols) elastic moduli of SiO_2, GeO_2, and B_2O_3 as a function of temperature. Arrows indicate the glass transition temperatures.

viscosity coefficients are measured by this method. Measurements of viscoelastic properties are done without mechanical contact. Properties determined by this technique characterize the equilibrium state of the material because they are probed by thermal vibrations that exist naturally within the material. Because no external shear rate is applied, the viscosity coefficient represents the zero-shear rate limit. And, given the short periods between compression and expansion in the sound waves probed by this experiment, the atomic-scale mechanisms underlying deformation and viscous dissipation are detected by this technique.

The unusual feature revealed by the data in Fig. 1 is the positive temperature dependence of the elastic moduli. Both SiO_2 and GeO_2 show this behavior for the glass between room temperature and T_g. In the case of SiO_2, the positive slope continues through the glass transition into the supercooled liquid. In the case of GeO_2, the slope sharply turns negative at T_g, but quickly resumes a positive slope. The data for B_2O_3 shows little temperature dependence below T_g. There is a cusplike increase near T_g, and then the elastic moduli abruptly drop. A minimum is reached near 700°C and then the moduli keep increasing for the remainder of the experimentally covered temperature range. The decrease in elastic storage capacity of B_2O_3 immediately above T_g can be attributed to the structural changes that take place in the supercooled melt, and which in fact constitute the transformation between glass and liquid (or vice versa). Once in the liquid state, the elastic moduli of B_2O_3 increase again.

The fact that B_2O_3 melt exhibits an elastic restoring force to shear deformation at 1000° above the melting temperature of the crystalline phase is astonishing in itself. At first glance one might hold the high probing frequencies responsible for this fact and classify the observed behavior as viscoelastic relaxation. Accordingly, while a viscoelastic medium flows in response to applied constraint at low frequencies, the higher the deformation frequency, the stiffer it acts. The added stiffness is apparent and due to kinematic reasons; it is referred to as relaxational modulus. Similarly, for a constant deformation frequency the medium flows at high temperatures, where structural relaxation occurs quickly in comparison to the deformation period, and stiffens toward lower temperatures when the atomic mobilities responsible for relaxation slow down. This concept is in fact a popular working model for describing the mechanical behavior of glass-forming liquids through the glass transition. However, as we showed in a series of papers, this approach does not at all reflect the actual atomic-scale processes that underlie the glass transition.[18–21] Furthermore, the concept of relaxational modulus does not explain the increase in the elastic modulus of the liquid with increasing temperature.

Studies of the mechanical properties of glasses under high pressure (i.e., in excess of several GPa) are less common, because under these conditions

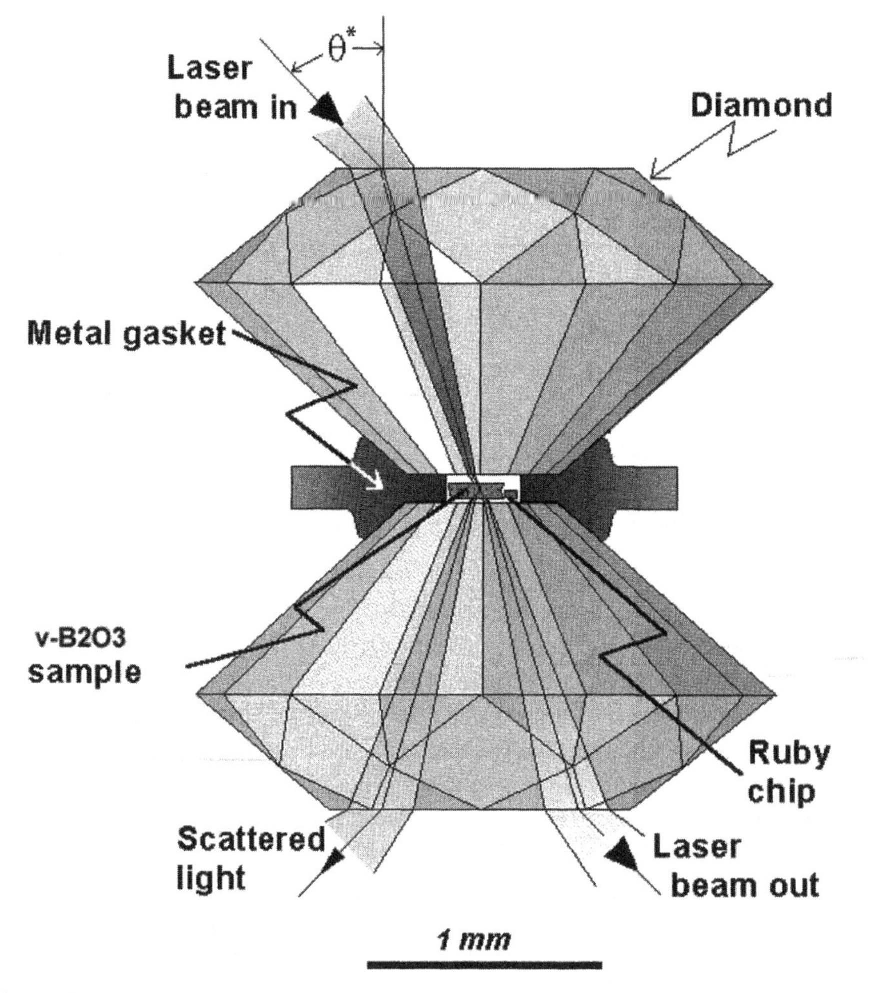

Figure 2. Schematic of a Merrill-Basset diamond anvil cell for high-pressure light scattering experiments.

it is difficult to establish contact with the appropriate probes. Again, Brillouin scattering provides an elegant way for measuring elastic constants.[22] A schematic of the experimental setup is shown in Fig. 2. For high-pressure experiments, the sample is confined between the flattened tips of two diamonds. A metal gasket prevents the sample material from being squeezed out from between the diamond surfaces. A small hole in the gasket (~0.25 mm diameter) forms the sample chamber. The incident laser beam and the

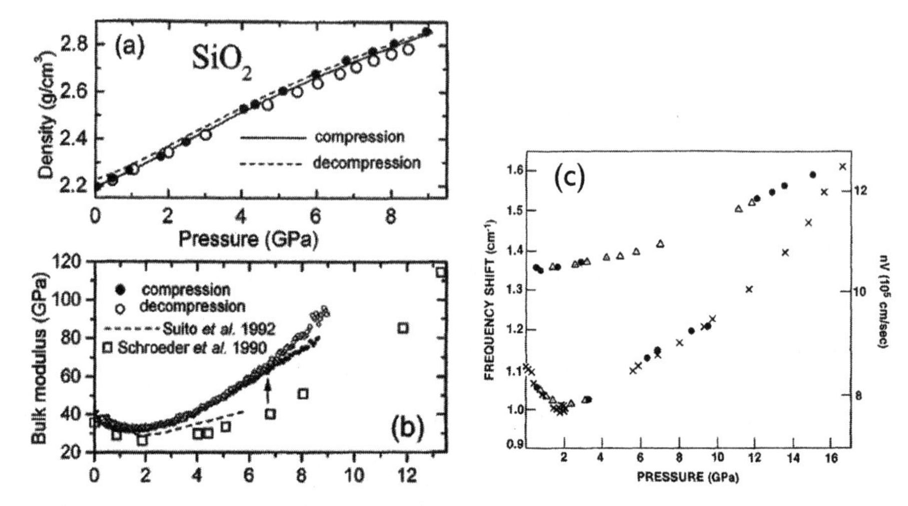

Figure 3. Anomalous pressure dependence of the mechanical properties of silica glass. Data are by (a) and (b) Tsiok et al.[10] and (c) Grimsditch et al.[11]

scattered light pass through the diamonds. The pressure inside the sample chamber is determined from the frequency shifts in the fluorescence spectrum of tiny ruby crystals that have been enclosed alongside the specimen. Pressures up to 100 GPa can be reached with this technique.

Figure 3 shows the pressure dependence of the elastic properties of silica glass obtained using Brillouin scattering and, for comparison at the low-pressure regime, using a strain-gauge technique. The anomalous behavior, which is reflected in the initial decrease of the bulk modulus, is evident in all data sets. From Fig. 1 we know that in response to temperature changes, anomalous behaviors are common to all strong glass formers. In response to pressure changes, anomalous behavior is most prominent in SiO_2, less so in GeO_2, and absent in B_2O_3. In the following section we will therefore explore the nature of this behavior by focusing on SiO_2. However, we anticipate that a similar explanation holds for the other network glass formers.

Atomic-Scale Simulations

The numerical technique we used to investigate the mechanisms that underlie the aforementioned anomalous properties of silica glass is called molecular dynamics (MD) simulation. This method consists of integrating the equations of motion of a collection of atomic-sized particles that interact with one another as described to a suitable potential function. The negative derivative of this potential with respect to the interatomic distance yields the force between two atoms, and the total force acting on a given atom is the sum of all pair-wise interactions involving this atom. Once this force is determined each atom is allowed to advance in direction of the total force acting on it, according to Newton's equation of motion. The duration of this motion is very short, only a few femtoseconds, because as soon as the atomic positions change, the forces on the atoms change too and they need to be re-evaluated. Hence, the atomic trajectories are solved for iteratively, by repeating the same sequence of force evaluations and particle accelerations millions of times. The result, albeit for a short duration compared to experimental time scales, yields a detailed account of atomic vibrations, transport processes, and equilibrium configurations. With today's computers one can follow the trajectories of configurations with millions of atoms. From the structure and dynamics of molecular configurations, it is possible to compute any desired property.

The reliability of MD simulations depends on the realism with which atomic interactions are modeled. Hence, researchers spent significant effort to develop such models. We have created a model that is particularly suitable for the simulation of crystalline and amorphous silica (among other inorganic compounds). Details about our interaction potential may be found in our previous publications.[23–25] We base the trust in the reliability of our interaction model on the many materials characteristics that we are able to reproduce in simulations of silica. For example, we can simulate various silica polymorphs without having to change any parameters in the interaction model. We can therefore study transformations between such polymorphs. If we start a simulation with the structure of α-cristobalite, we obtain β-cristobalite upon heating or by applying a hydrostatic tensile stress at room temperature. The X-ray diffraction pattern of the simulated β-cristobalite structure matches the experimental one. Furthermore, the simulated infrared spectra of cristobalite match the experimentally observed spectra and exhibit the same changes upon varying the temperature (i.e., changes that are the result of the transformation between α- and β-cristobalite). Based on our simulations we were able to resolve a long-standing controversy concerning the true structure of β-cristobalite.

Nature of the Anomalous Thermomechanical Properties in Silica Glass

To understand the anomalous behavior of silica glass, it is useful to review a few details about the structure and properties of α- and β-cristobalite, and the transformation between these two modifications. In cristobalite, similar

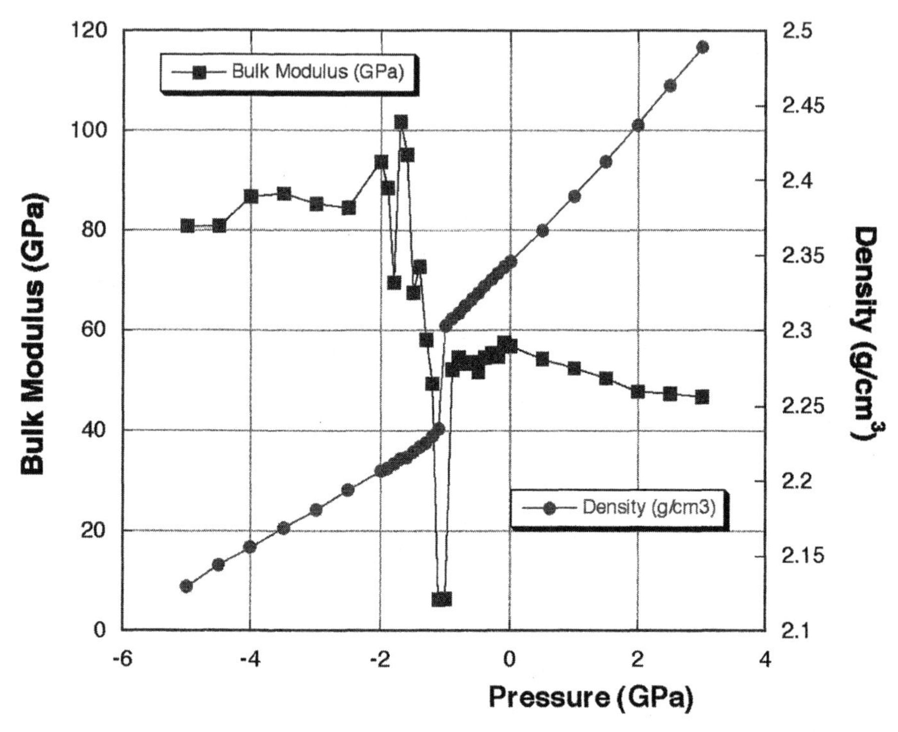

Figure 4. Bulk modulus (solid squares) and density (open circles) of simulated cristobalite silica as a function of the pressure. To the left of the discontinuity β-cristobalite, which has the lower density but higher modulus, is stable and to the right α-cristobalite, which has the higher density and lower modulus.

to the case in many other silica polymorphs, silicon is tetrahedrally coordinated by oxygen, and by sharing oxygens between tetrahedra these units are linked into continuous network structures. The resulting Si-O-Si links form an angle of approximately 150° because of the lone pair of electrons located on the oxygen. In both α- and β-cristobalite only one type of closed-loop structure can be identified within the network of tetrahedra, namely six-membered rings. Hence, the transformation between the two polymorphs can occur with all network bonds remaining intact. We determined that the mechanisms by which this transformation takes place is a cooperative rotation of Si-O-Si bonds by 90° around the Si-Si axis. If the transformation is induced by pressure changes, this rotation is abrupt and definitive, whereas if it is thermally induced not all Si-O-Si bonds rotate in unison; a significant degree of disorder is induced in the structure and only on a time average are the bonds rotated by 90°.

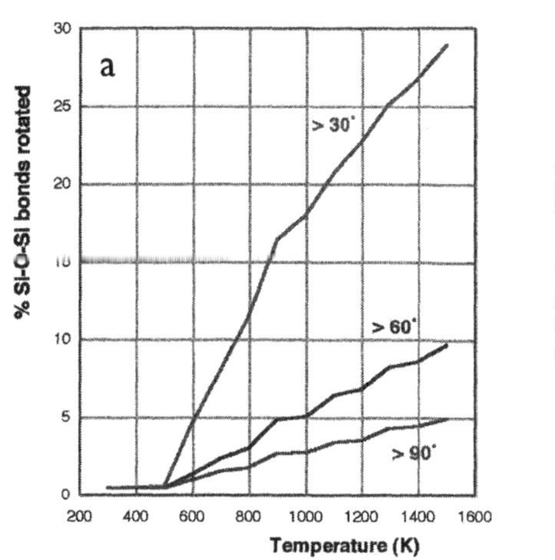

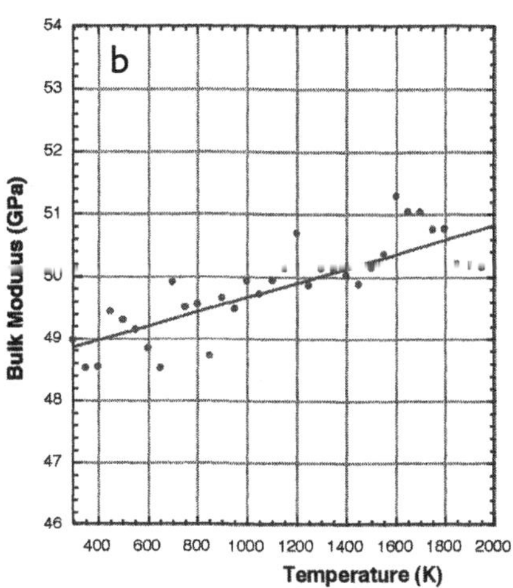

Figure 5. (a) Fraction of Si-O-Si bonds that have rotated by more than 30°, 60°, or 90° relative to their room temperature orientation. (b) Bulk elastic modulus of simulated silica glass.

Figure 4 shows the density and the bulk modulus of cristobalite as a function of pressure through the α-to-β transition. At the transition pressure of –1 GPa, we observe a step-change in the density, as expected for a first-order transition, and the bulk modulus dips down to nearly zero as a result of mode softening associated with the transformation. The interesting point to note is that, when going from the high- to the low-density modification, the elastic modulus increases. This behavior is observed experimentally and we reproduce it in our simulations. It is due to the fact that in the low-density β-cristobalite the geometry of an individual network ring is highly symmetric and all Si-O-Si bridges point outward, whereas in α-cristobalite rings some bridges point inward because of a partial twist in the ring structure. In β-cristobalite applied stresses result in straining bonds between atoms or bending the angles formed by two adjoined bonds, whereas in α-cristobalite some of the load can be absorbed by advancing the twist deformation. This latter mode of deformation is responsible for the softer elastic response of α-cristobalite.

Our simulations revealed that Si-O-Si bond rotations, very similar to the ones underlying the α-to-β transformation in cristobalite, also take place in amorphous silica. Fig. 5(a) shows the fraction of Si-O-Si bonds that have rotated by more than a given angle relative to their orientation in the room-temperature glass, as a function of temperature. Note that these quantities

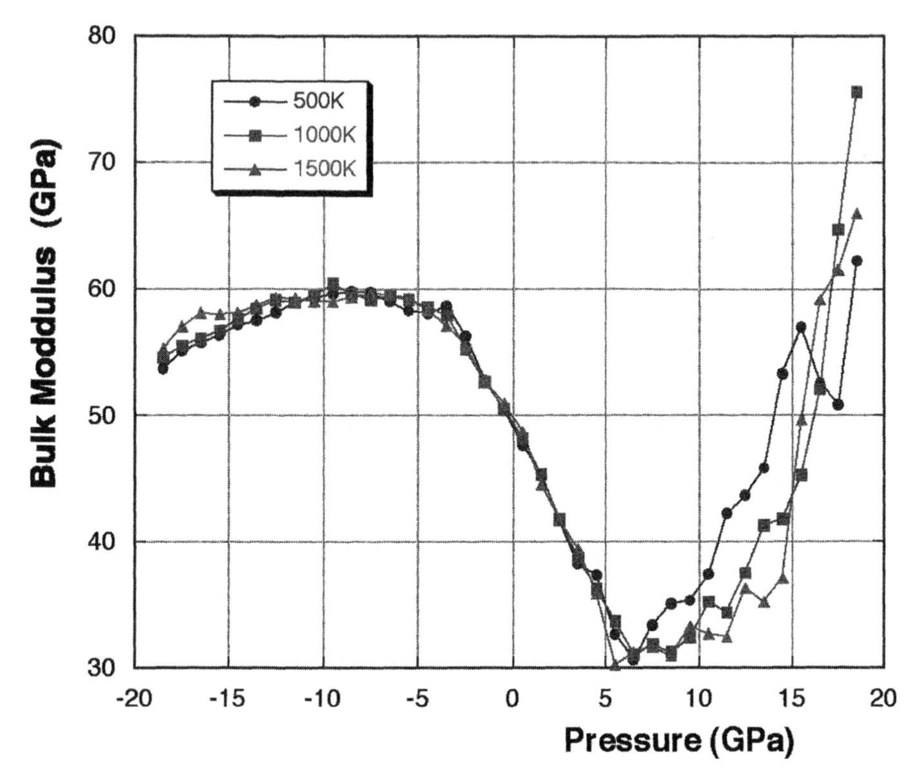

Figure 6. Bulk modulus of simulated amorphous silica as a function of pressure. Between –9 and +7 GPa, where the pressure dependence of the modulus is negative, corresponds to the regime of anomalous behavior. Beyond this pressure range the behavior is normal.

represent a dynamic equilibrium and correspond to the average over the entire simulated configuration at any given moment in time. The extent of rotation changes from one bond to another, and for any given bond it also changes over time. Fig. 5(b) shows the corresponding change in elastic modulus of the simulated glass, which reproduces well the experimentally observed positive temperature dependence.

Hence, we believe that in the room temperature glass network rings are arranged in a fashion similar to α-cristobalite; that is, they have concave geometries in order to provide for dense atomic packing. When the glass is heated these rings inflate because of thermal motion and locally transform into geometries closer to that in β-cristobalite. This change to more convex ring geometries with increasing temperature is the reason for an overall increase in elastic modulus.

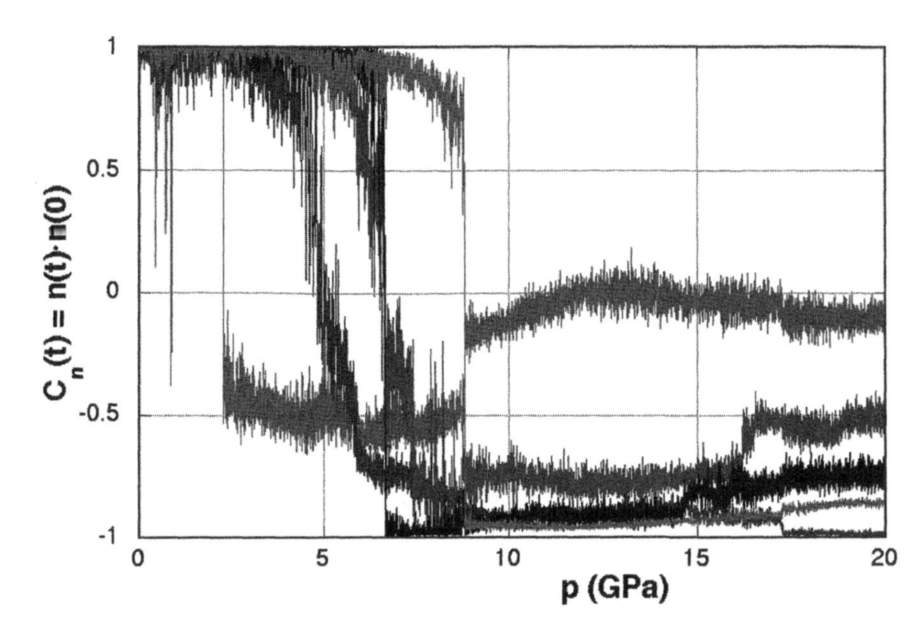

Figure 7. Orientation of individual Si-O-Si bond planes as a function of pressure, expressed in terms of the dot product between the current unit normal vector and that of a reference configuration (zero pressure starting configuration in this case).

Ultimately, the elastic modulus of amorphous silica increases with temperature because the glass has retained the ability to undergo structural transitions similar to those that occur in its crystalline counterparts. The similarity of the underlying mechanisms becomes even more obvious when examining the pressure dependence of the elastic properties of silica glass. Figure 6 shows the bulk modulus of simulated silica glass as a function of pressure at three different temperatures. The minimum occurs at slightly higher pressures compared to experiments, probably because the simulated glass has a significantly higher fictive temperature, but the anomalous decrease in elastic modulus with increasing pressure is well reproduced. In simulations we can also easily establish conditions of tensile hydrostatic stress, and we found that the anomalous behavior continues into tensile regime (which corresponds to the deformation resulting from thermal expansion) until a modulus maximum is reached at about –9 GPa.

Figure 7 shows that indeed structural changes similar to those underlying the α-to-β transformation in cristobalite are responsible for this anomalous behavior. In this graph, a measure of the orientation of Si-O-Si bonds is plotted as a function of pressure. Because the two Si-O bonds connecting a given oxygen form an angle less than 180°, these two bonds define a plane and we can use orientation of the normal to this plane as a measure for the degree of rotation. We compare the orientation of a plane with that at zero pressure, as a reference. Figure 7 in essence shows the cosine of the angle of rotation of four individual planes, for as long as this measure is close to 1, the bond plane has not rotated. A value of zero means 90° rotation and –1 means 180° rotation. The important information revealed by this graph is that individual bonds rotate suddenly and that different bonds rotate at different degrees of densification.

Because of the structural disorder, stresses transmitted by amorphous structures do not affect every spatial region to the same extent. Stresses concentrate in locations that originally formed with pronounced departures from minimum energy (or relaxed) configurations, that is, in locations with inherent defects. These high–stress intensity regions are the first to respond to the imposed constraints with a localized structural transition. Subsequent to such a transition, the surrounding structure can relax and it takes additional deformation before the next transition is invoked. Each transformed region contributes a different modulus to the composite materials property. The abrupt nature of individual rotations demonstrates that the corresponding structural changes are indeed transitions and not gradual deformations. Only because of the cumulative effect of each transition do the changes in elastic properties appear to occur gradually from a macroscopic point of view.

Conclusions

We have discussed two techniques of investigation that may not currently be commonplace in the repertoire of glass manufacturing research and development but that may be of interest for materials design and process control: Brillouin light scattering and molecular dynamics computer simulations. Both techniques have the advantage of allowing for investigation of materials under extreme temperature and pressure conditions, and provide insight into the molecular-scale processes responsible for materials behaviors. We have used both methods in a synergistic investigation of the anomalous thermomechanical properties of silica glass. We have presented evidence that both the anomalous dependences of the elastic modulus on temperature and pressure are due to localized structural transitions similar to those observed in the α-to-β cristobalite phase transformation. Accordingly, at high densities and low temperatures the network ring structure of the glass is collapsed. This provides for high packing density but low elastic modulus. Upon expansion of the structure, through heating or in response to tensile loading, network rings inflate to a geometry that is more resilient to deformation.

Acknowledgements

We would like to acknowledge financial support by the National Science Foundation under grant no. DMR-0230662.

References

1. J. G. Collins and G. K. White, *Prog. Low. Temp. Phys.,* **4**, 450 (1964).
2. A. Polian, D. Vo-Thanh, and P. Richet, "Elastic Properties of α-SiO_2 up to 2300 K from Brillouin Scattering Measurements," *Europhys. Lett.,* **57**, 375 (2002).
3. M. Fukuhara and A. Sanpei, "High Temperature-Elastic Moduli and Internal Dilational and Shear Frictions of Fused Quartz" *Jpn. J. Appl. Phys.,* **33**, 2890 (1994).
4. R. E. Youngman, J. Kieffer, J. D. Bass, and L. Duffrène, "Extended Structural Integrity in Network Glasses and Melts," *J. Non-Cryst. Solids,* **222**, 190 (1997).
5. P. W. Bridgman, *Am. J. Sci.,* **10**, 359 (1925).
6. P. W. Bridgman, *Proc. Am. Acad. Arts Sci.,* **76**, 9 (1945).
7. P. W. Bridgman, *Proc. Am. Acad. Arts Sci.,* **76**, 71 (1948).
8. C. Meade and R. Jeanloz, *Phys. Rev. B,* **35**, 236 (1987).
9. M. R. Vukevich, *J. Non-Cryst. Solids,* **11**, 25 (1972).
10. O. B. Tsiok, V. V. Brazhkin, A. G. Lyapin, and L. G. Khvostantsev, "Logarithmic Kinetics of the Amorphous-Amorphous Transformations in SiO_2 and GeO_2 Glasses under High Pressure," *Phys. Rev. Lett.,* **80**, 999 (1998).
11. M. Grimsditch, "Polymorphism in Amorphous SiO_2," *Phys. Rev. Lett.,* **52**, 2379 (1984).
12. M. Grimsditch, "Annealing and Relaxation in the High-Pressure Phase of Amorphous SiO_2," *Phys. Rev. B,* **34**, 4372 (1986).
13. M. Grimsditch, R. Bhadra, and Y. Meng, "Brillouin Scattering from Amorphous Materials at High Pressures," *Phys. Rev. B,* **38**, 7836 (1988).
14. R. J. Hemley, H. K. Mao, P. M. Bell, and B. O. Mysen, "Raman Spectroscopy of SiO_2 Glass at High Pressure," *Phys. Rev. Lett.,* **57**, 747 (1986).
15. C. Meade, R. J. Hemley, and H. K. Mao, "High-Pressure X-Ray Diffraction of SiO_2 Glass," *Phys. Rev. Lett.,* **69**, 1387 (1992).
16. S. Susman, K. J. Volin, D. L. Price, and M. Grimsditch, "Intermediate-Range Order in Permanently Densified Vitreous SiO_2: A Neutron-Diffraction and Molecular-Dynamics Study," *Phys. Rev. B,* **43**, 1194 (1991).
17. R. E. Youngman, J. Kieffer, J. D. Bass, and L. Duffrène, "Extended Structural Integrity in Network Glasses and Melts," *J. Non-Cryst. Solids,* **222**, 190 (1997).
18. J. Kieffer, J. E. Masnik, O. Nickolayev, and J. D. Bass, "Brillouin Scattering Investigation of Alkali Tellurite Melts," *Phys. Rev. B,* **58**, 694 (1998).

19. J. E. Masnik, J. Kieffer, and J. D. Bass, "The Complex Mechanical Modulus as a Structural Probe: Case of Alkali Borate Liquids and Glasses," *J. Chem. Phys.,* **103**, 9907 (1995).
20. J. Kieffer, "Structural Transitions in Glasses and Glass-Forming Liquids (Morey Award Lecture)," *Am. Ceram. Soc. Bull.,* **81**, 73 (2002).
21. J. Kieffer, "Structural Transitions and Polyamorphism in Glass-Forming Oxides," *J. Non-Cryst. Solids,* **307–310**, 644 (2002).
22. J. Nicholas, R. Youngman, S. Sinogeikin, J. Bass, and J. Kieffer, "Structural Changes in Vitreous Boron Oxide," *Phys. Chem. Glasses,* **44**, 249 (2003).
23. T. Nanba, J. Kieffer, and Y. Miura, "Molecular Dynamic Simulation of the Structure of Sodium-Germanate Glasses," *J. Non-Cryst. Solids,* **277**, 188 (2000).
24. L. Huang and J. Kieffer, "Molecular Dynamics Study of Cristobalite Silica Using a Charge Transfer Three-Body Potential Model," *J. Chem. Phys.,* **118**, 1487 (2003).
25. D. C. Anderson, J. Kieffer, and S. Klarsfeld, "Molecular Dynamic Simulations of the Infrared Dielectric Response of Silica Structures," *J. Chem. Phys.,* **98**, 8978 (1993).

Energy and Combustion

Advanced Investigation Methods for the Characterization of Flames Aimed at an Optimization of the Heat Transfer Processes in Glass Melting Furnaces

Axel Scherello
Gaswärme-Institute.V., Essen, Germany

Introduction

Because flames have to be adjusted to the combustion area of each glass melting furnace, it is important to be able to describe and characterize the flames. This can be done by the following five properties: flow, mixture, fuel conversion, heat release, and pollutant formation. These properties are listed in order of their influence on each other. The flow field influences the mixture of fuel and oxygen, and the mixture is responsible for the combustion process itself, which affects the heat release of the flame. Arranged at the end of this chain stands pollutant formation, which is influenced by most of the other flame properties but does not have a significant influence on the others.

In order to describe the flames and to optimize their above-listed properties, advanced investigation methods can be applied. This paper describes nonstandard measurement techniques for velocity, mixture, and combustion detection, for example, particle image velocimetry, mixture characterization by laser light sheet visualization, and combustion species measurement by laser-induced fluorescence. In addition, heat transfer probes were introduced, which help to judge the efficiency of the heat transfer from the flames toward the crown and into the glass. Most of the techniques described here can be applied not only in laboratory-scale furnaces but in production furnaces in order to realize in situ optimization of the glass melting process.

Velocimetry

The basis of all combustion phenomena is flow, which is necessary to transport fuel and oxidant into the furnace. The velocity distribution influences the mixture and therefore combustion and heat release. But strong feedback from the other phenomena affect the velocity of the furnace-entering jets. The most important influence is the heat release of the burner bricks toward

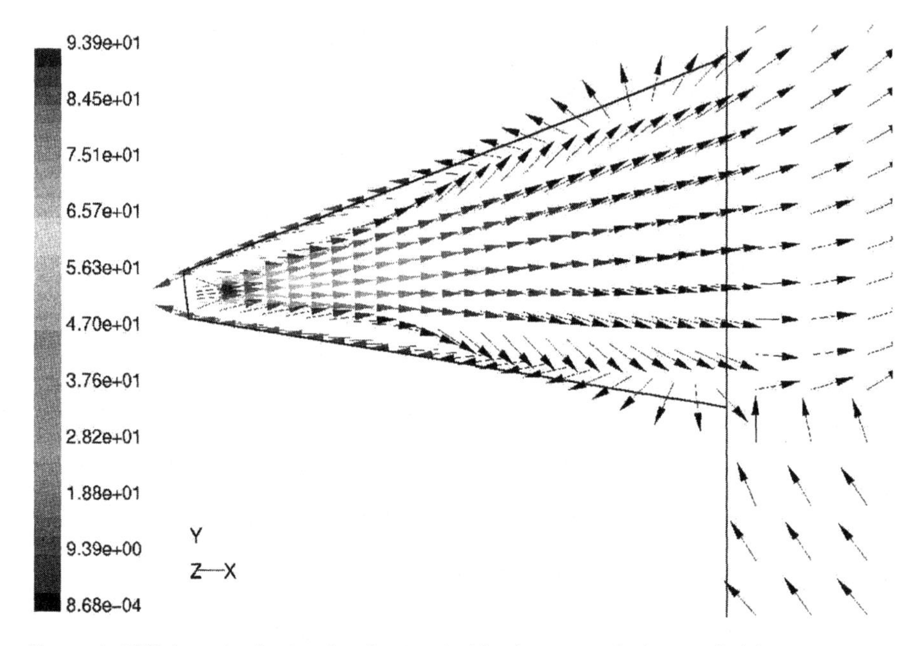

Figure 1. CFD-based velocity distribution inside the cone of a burner brick.

natural gas flows. Combined with a mixture of natural gas and recirculated flue gases, the heatup of fuel leads to cracking and recombination of the hydrocarbons, which results into a strong increase of soot formation. This temporarily formed soot is burned within the main reaction zone of the flame. In addition, soot is a solid body and therefore a more effective radiator than the gaseous flame products CO_2 and H_2O. Because of an increase of soot formation, a brighter shining flame is produced.

Figure 1 shows the numerical simulation of the velocity distribution inside a burner brick. There is a clear backward directed flow along the cones wall toward the root of the fuel jet. This backward-streaming volume flux is heated up at the brick walls and then mixed into the natural gas jet.

Unfortunately there is no possibility of measuring velocities inside burner bricks. Pressure drop–based probes cannot be applied into glass furnaces or high-temperature test furnaces. Advanced velocity measurement devices such as laser doppler anemometry (LDA) and particle image velocimetry (PIV) are based on detection rectangular toward the velocity direction. But because of cold burner brick model laboratory experiments and PIV results in front of the burner bricks, computational fluid dynamics (CFD) produces reliable results.

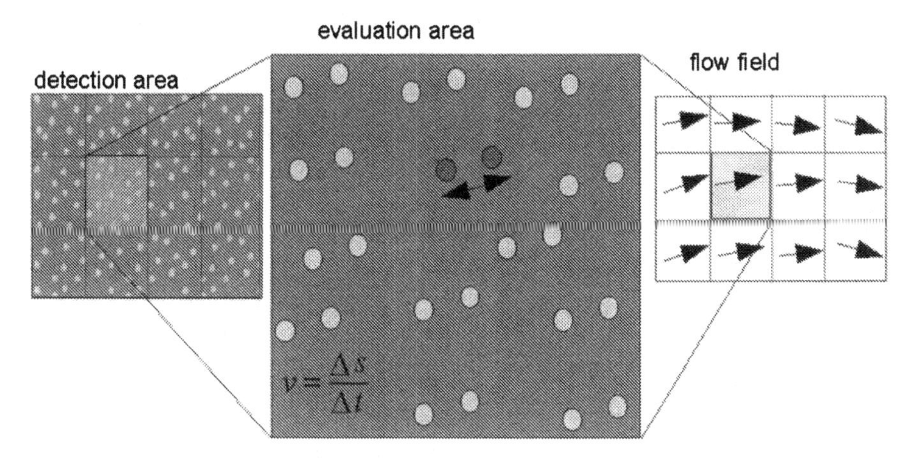

Figure 2. Detection principle of particle image velocimetry.

Particle image velocimetry, an advanced velocity meaurement method, has become an important tool for the quantitative and nonintrusive investigation of flow phenomena. The PIV principle is shown in Fig. 2. For measurements the flow is seeded with tiny particles called "tracers" — for example, oil or water aerosols in air and solid particles in fluids or flames. Using optics the beam of a double-pulse laser is formed to a light sheet. This plane is placed through the flow twice within a short time interval, and the tracers are illuminated. The light scattered by the particles is recorded in a double frame of a CCD camera. The observation position of the camera is usually placed perpendicular to the direction of illumination. Depending on the flow velocity and the factor of magnification of the camera lens, the delay of the two pulses must be chosen such that adequate displacements of the particle images on the CCD are obtained. From the time delay between the two illuminations and the displacement of the tracer, velocity vectors can be calculated. (Detailed information can be found in Refs. 1–3.)

Figure 3 shows the application of PIV in a test furnace at GWI laboratories. The left picture shows the laser pathway into the furnace. The right picture shows the laser light inside of the furnace and the tip of the tested burner.

Figure 4 shows the results of this PIV experiment. On the left side are positioned two corresponding single images that resulted in the flow field of an adjustable FLOX burner* shown on the right side.

*FLOX is a registered trademark of WS Wärmeprozesstechnik, Renningen, Germany.

Figure 3. Left: Laser beam path and light sheet optics. Right: Laser light sheet inside the test furnace.

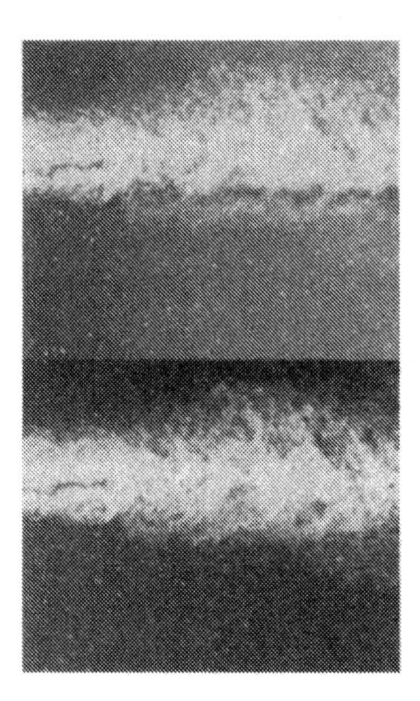

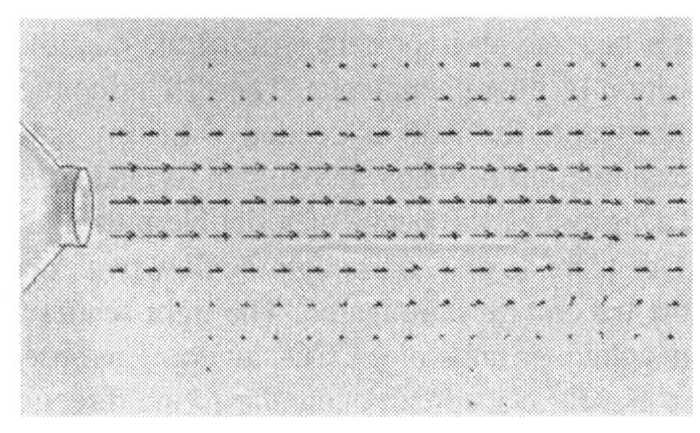

Figure 4. Corresponding double images (left) and evaluated flow field (right).

Mixture

In order to detect the mixture of fuel and oxidant, classical methods such as local suction pyrometry and gas analysis are applied. Another method is laser light sheet visualization (LSV). In the same way it was realized in the case of PIV, a laser beam is led into the furnace and widened to a laser light

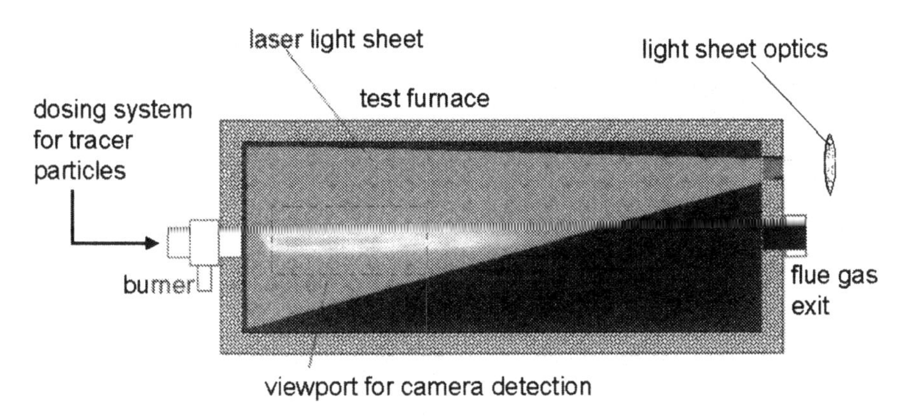

Figure 5. Setup of the LSV system.

sheet using a cylindrical optic. The light sheet is positioned vertically through the center of the burner. (See Fig. 5.)

The fuel volume flux is seeded with zircon oxide particles sized from 0.3 up to 2.0 μm that stay solid up to 2980°C. The laser light is scattered at these tracer particles. The Mie-scattered light is detected with a CCD camera whose exposure times are triggered parallel to the laser frequency.

The images are transferred to a computer where data acquisition takes place for about 100 frames. The measured scattering light intensity is proportional to the quantity of particles at a special discrete volume. The particle quantity/scattered light intensity at the fuel nozzle (equivalent to 100% natural gas) is the reference value. A data evaluation results in the mixture fraction field and its standard deviation distribution. (Detailed information can be found in Refs. 4 and 5; see Fig. 6.)

Fuel Conversion

In order to characterize the combustion process itself, the classical methods use indirect procedures such as temperature measurements or water-cooled probe–based gas analysis. Both methods are only spot measurements, which cannot answer the question of flame length and width. Figure 7 shows two images taken in glass melting furnaces. In both cases only the soot radiation of these natural gas flames can be seen because of its higher radiation intensity compared with the radiation of the gaseous flame products.

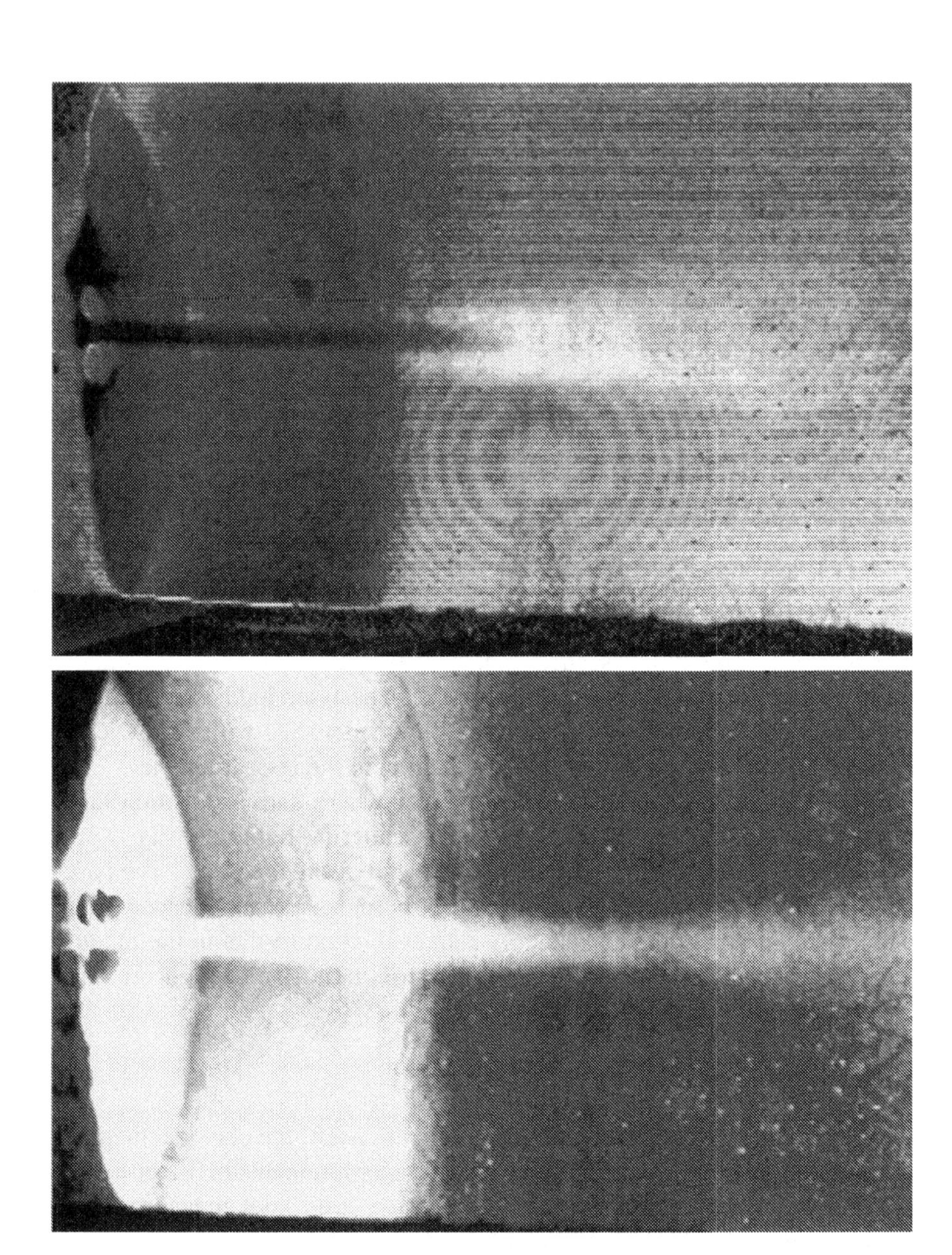

Figure 6. LSV results: Mixture fraction distribution of a coaxial gas/air injection burner (top) and concentration fluctuation distribution (bottom).

Figure 7. Observation camera images from inside a cross-fired (top) and an end-fired glass-melting furnace.

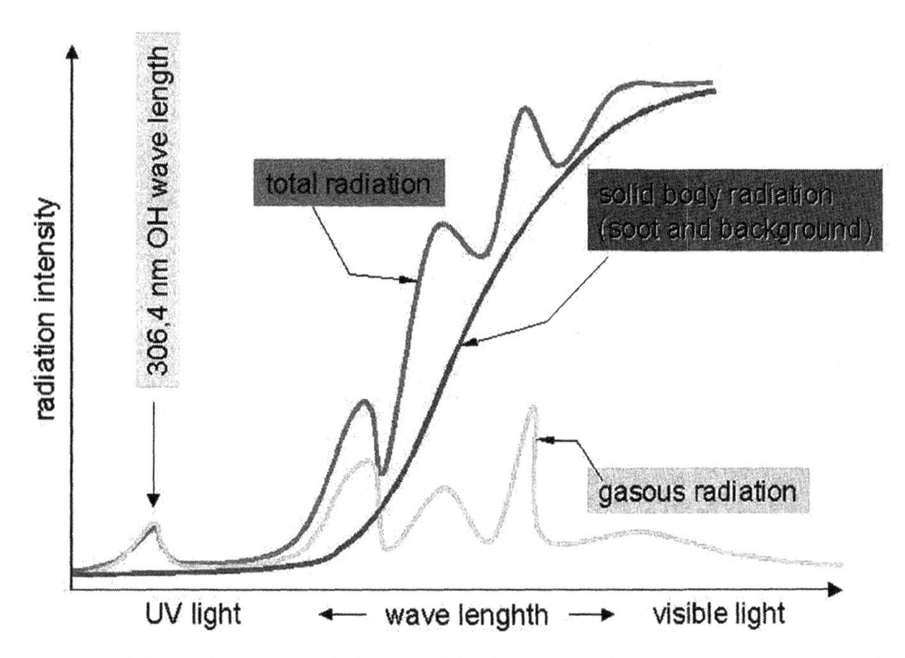

Figure 8. Schematic diagram of flame and background radiation intensity versus their wavelengths.

UV Flame Visualization

Following the diagram shown in Fig. 8, a solution for wiping out the bright shining background is to detect the flames not over a wide spectral range but only at a specific wavelength. A suitable way is to use a radiation band around 306.4 nm where only the combustion radical OH emits its typical radiation.[6,7] Using a light-intensified CCD camera system together with a small band UV filter (Fig. 9), it is possible to detect the distribution of the combustion radical OH.

OH radicals are intermediate reaction products formed by high-temperature dissociation effects. Because OH is mainly responsible for the oxidation of CO the application of this UV visualization technique allows us to describe the contours of a flame. Additionally, the occurrence of high quantities of OH indicates local reaction intensity and temperature peaks. With this information it is possible to optimize combustion and to reduce pollutant emissions.

By means of several testing series we found that the intensity of OH radiation correlates with the NO_x emissions measured at the furnace exit. If the OH radiation intensity is bright, the measured NO_x emissions are high.

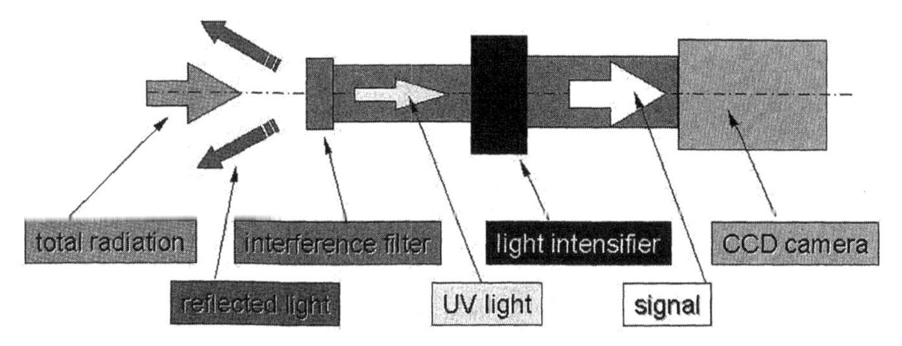

Figure 9. Detection principle of the UV flame visualization method.

In order to use this visualization and optimization tool for production furnaces, GWI and its partners have developed a UV endoscope-based system called Florian (Fig. 10). The cooling jacket–protected optic can be placed inside small furnace openings to allow sufficient flame observation.

Some results applying Florian are shown in Fig. 11. By visualizing the flames inside the precombustion chamber of a rotary furnace it was found that only three of four gas guns were in operation. This malfunction had not been detected before. Another example is the operation of an oxy-fuel burner at GWI test furnaces. Because of a change in the oxidation ratio, a strong increase in OH radiation was detected.

Laser-Based Methods

The OH visualization described above is not a quantitative method of fuel conversion detection. The three-dimensional body of a flame is visualized by detecting its natural OH radiation. In contrast to the OH radical visualization are laser-based methods where species radiation is forced by laser shots in a laser light sheet. These methods are planar quantitative species detection techniques that require an enormous extra effort in equipment and preparation.

OH-LIF Measurement

The laser-induced fluorescence (LIF) technique is another substantially more complex method of determining the extension of the reaction zone. In contrast to the OH radical visualization, the LIF technique excites electrons of OH radicals to a higher energy level using laser shots at wavelengths of 248 and 256 nm. After each laser shot, the energy level drops back to the original level and UV light is emitted at a shifted wavelength (~300 nm).

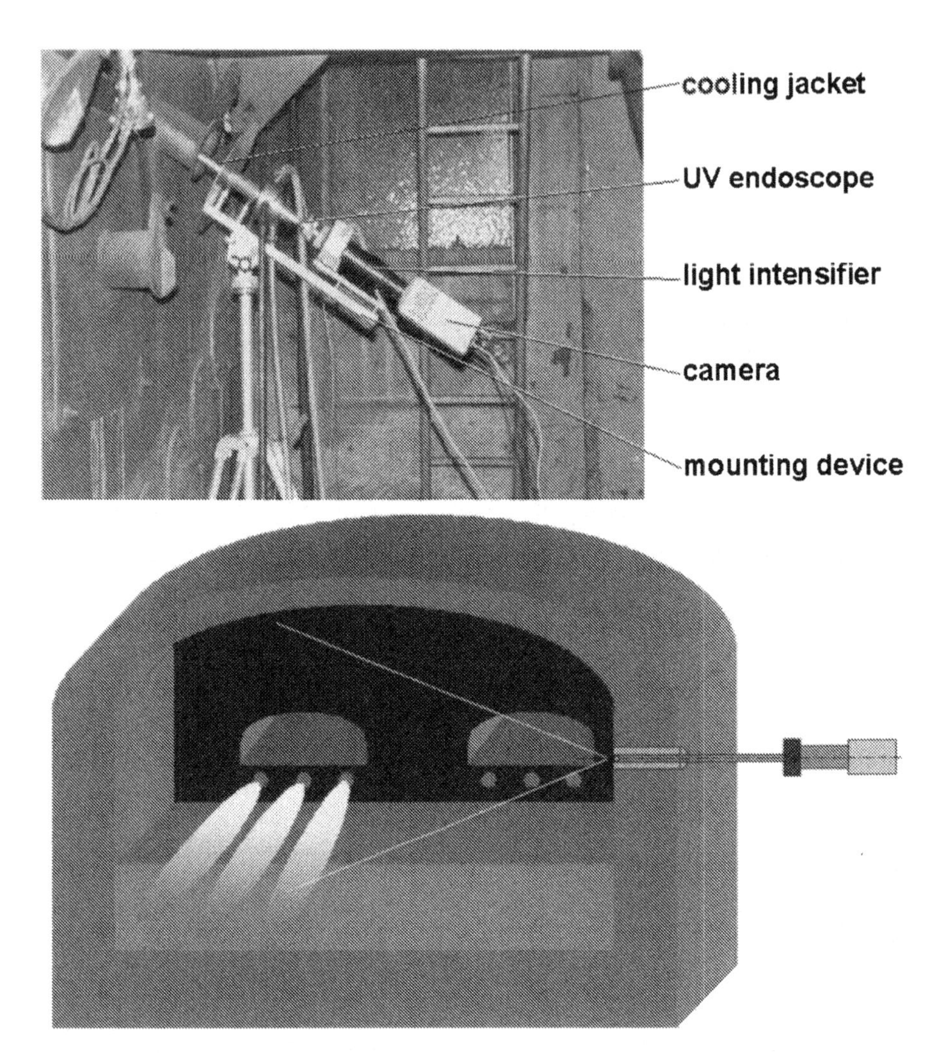

Figure 10. Application of Florian at a rotary furnace precombustion chamber (top) and a possible Florian positioning for a cross-fired glass furnace (bottom).

This UV light can be detected in the same way as in the OH visualization method using a light-intensified CCD camera system. A data evaluation provides quantitative OH distribution data. Using different exiting wavelengths other species (NO, O_2, CxHy, H_2, and H_2O) can be detected.

GWI operates an OH-LIF system at its combustion facilities. This system uses a tunable Nd-YAG laser that provides a laser wavelength of 256

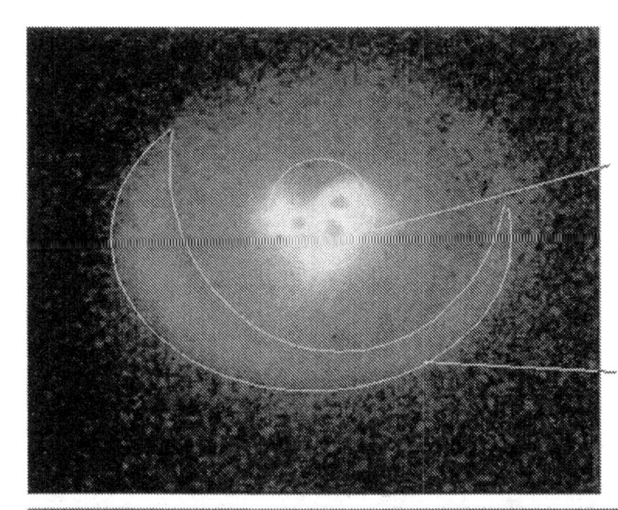

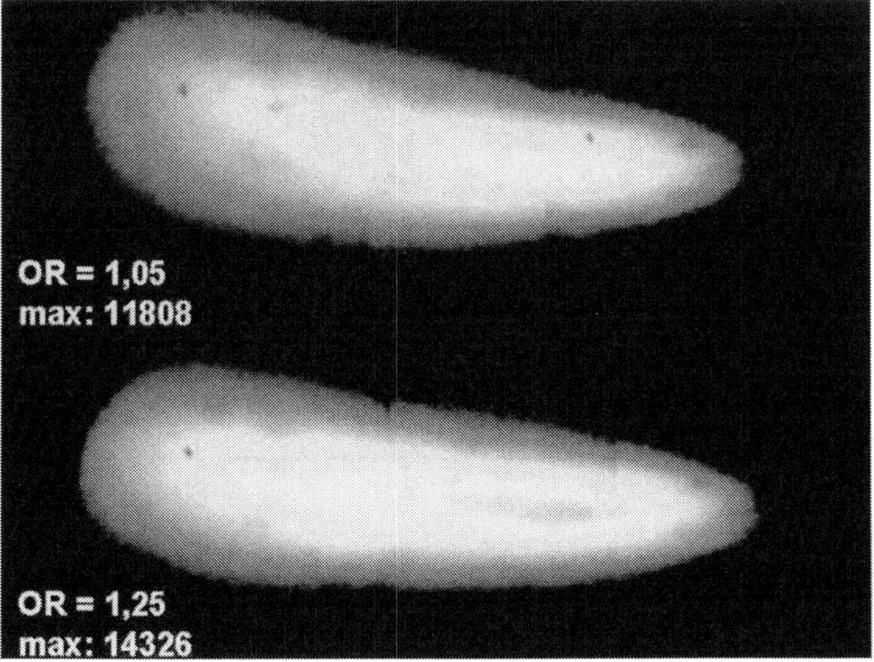

Figure 11. Frontal flame visualization of an 8-MW natural gas burner (top) and operation of an oxy-fuel burner at different oxidation ratios (bottom).

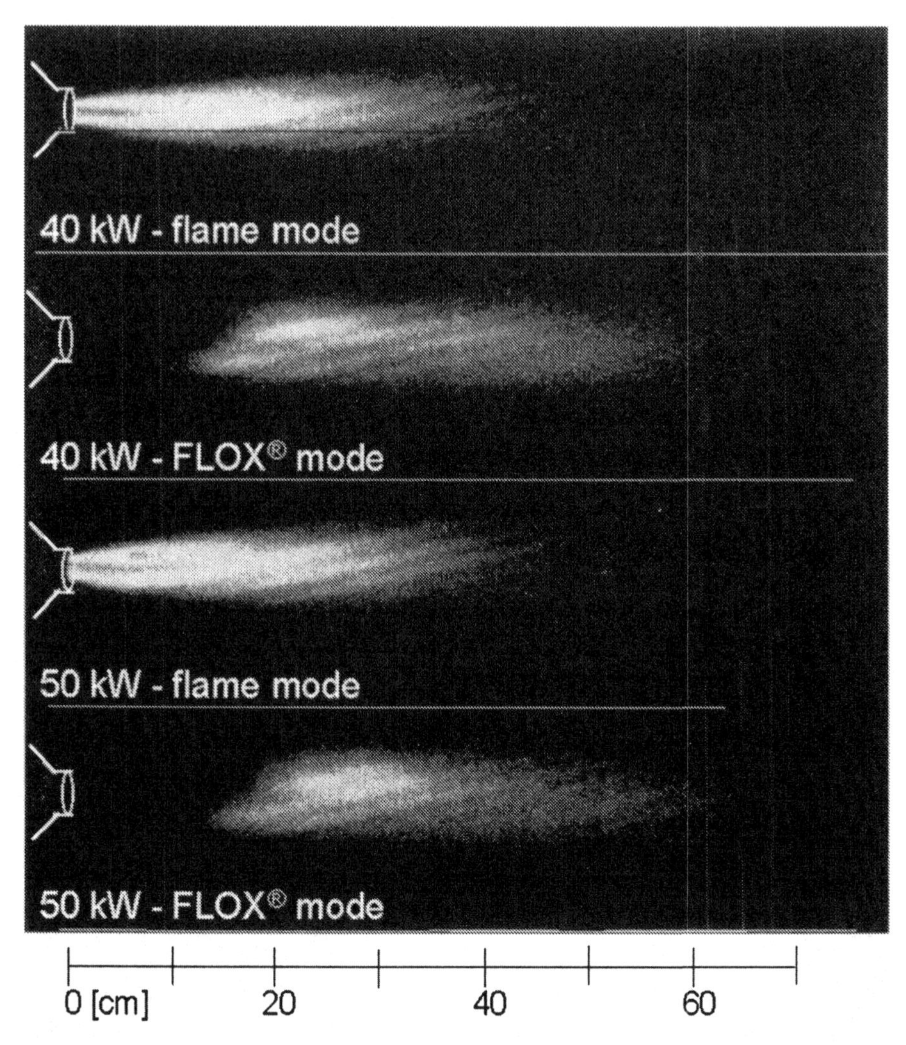

Figure 12. OH distribution for four different operation modes of a FLOX burner.

nm for OH excitement. Figure 12 shows the OH-LIF measurement results of the same burner, which serves for the PIV method as well.

In addition, Rayleigh or Raman imaging methods can be applied to characterize a combustion process. Rayleigh scattering can be used to measure planar temperature fields if the gas composition and the pressure are known. Using the inelastic Raman scattering process, the concentration of major combustion species such as N_2, O_2, CxHy, CO_2, H_2, and H_2O can be detected. For both techniques strong laser sources are required and fluorescence has to be avoided.

Figure 13. High-temperature test furnace (1650°C furnace wall temperature) and air preheater at GWI Laboratories.

Figure 13 shows an image of the high-temperature furnace at GWI laboratories. An air preheating device that provides 1250 mN^3/h air at 1350°C is shown on the left. The furnace is equipped with 250-mm diameter viewports that can be closed with double sliced, cooled, and purged windows [see Fig. 14(a)] in order to apply optical measurements.

During a research project aimed at the optimization of burner systems for glass tanks[8,9] GWI assigned Laser Laboratory (Göttingen), a well-known institute that specializes in laser measurements, to contractual work in order to obtain more detailed knowledge about the combustion processes of underport systems. During this laser measurement campaign it was discovered that the advantage of sooty flames regarding their radiative heat transfer is a disadvantage for the application of LIF and laser light scattering methods. Because of reflections at soot particles and the fluorescence of ring-shaped hydrocarbon molecules [Fig. 14(b)], no signals could be separated for evaluation.

Nowadays LIF as well as Raman and Rayleigh scattering techniques are applied not only for laboratory-scale combustion experiments but also for industrial-scale combustion facilities, such as gas turbines and engines.

Heat Release and Transfer

In order to evaluate the combustion process regarding its application in heating a product, the heat release and the heat transfer from flames toward the product must be detected. One method is to apply a heat-balancing procedure for the furnace. Because of several uncertainties about heat losses in

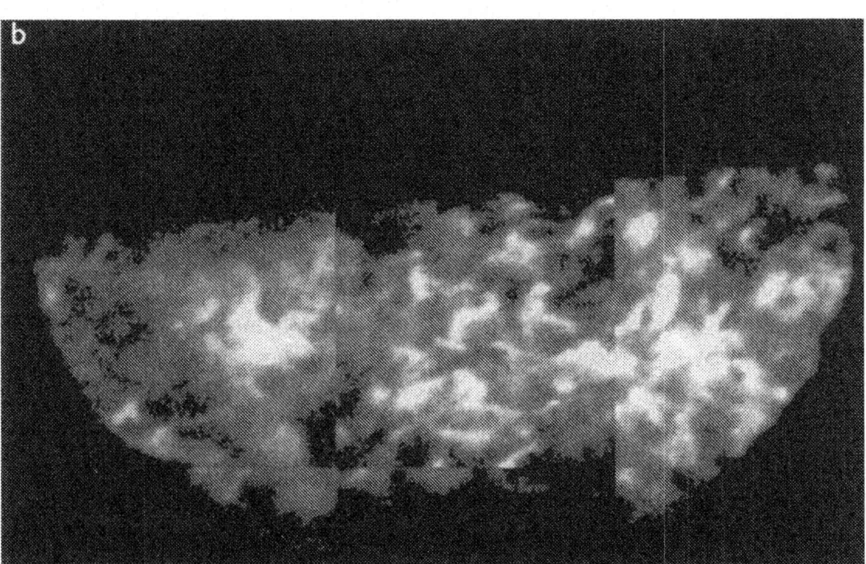

Figure 14. (a) Window covered furnace viewport during operation. (b) Scattered light from soot particles and fluorescence at C_xH_y molecules.

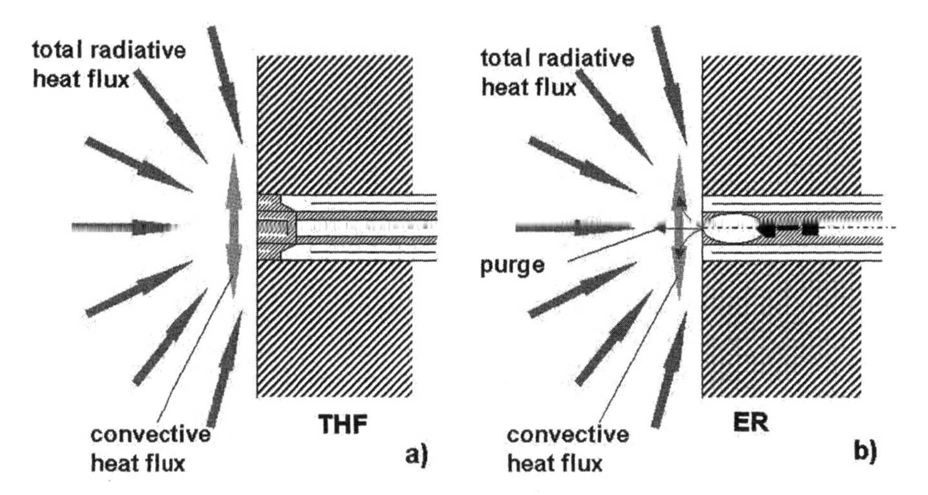

Figure 15. Detection principle of (a) the THF and (b) the ER probes positioned on the furnace wall level.

production furnaces, this method is not a very precise tool to assess flames properties.

Overall Heat Transfer Probes

Because the heat flux can be separated in conduction, convection, and radiation, GWI is operating different probes to measure these heat flux quantities. (Detailed information can be found in Refs. 10 and 11.)

The total heat flux probe [see Fig. 15(a)] detects the half-sphere total radiation toward the probe tips as well as the convective heat transferred from the hot atmosphere flowing along the walls.

The heat flux based on these two phenomena is transferred axially to the cooled rear side of the probe head. Because the conductivity of the head material is known, total heat flux and surface temperature can be determined by measuring temperatures at two points of the head. A sketch of the THF detector head is shown in Fig. 16.

In contrast to the THF, the ellipsoidal radiometer (ER) is equipped with a purging device that avoids the convective influence. The ER measures total radiation (wall radiation, flue gas radiation, and soot radiation) impinging upon the furnace wall from the furnace half-space at a given location [see Fig. 15(a)]. The radiation emitted by the furnace half-sphere (180° solid angle) enters the ellipsoid, which reflects and bundles the rays and focuses them on a point opposite the inlet (see Fig. 17).

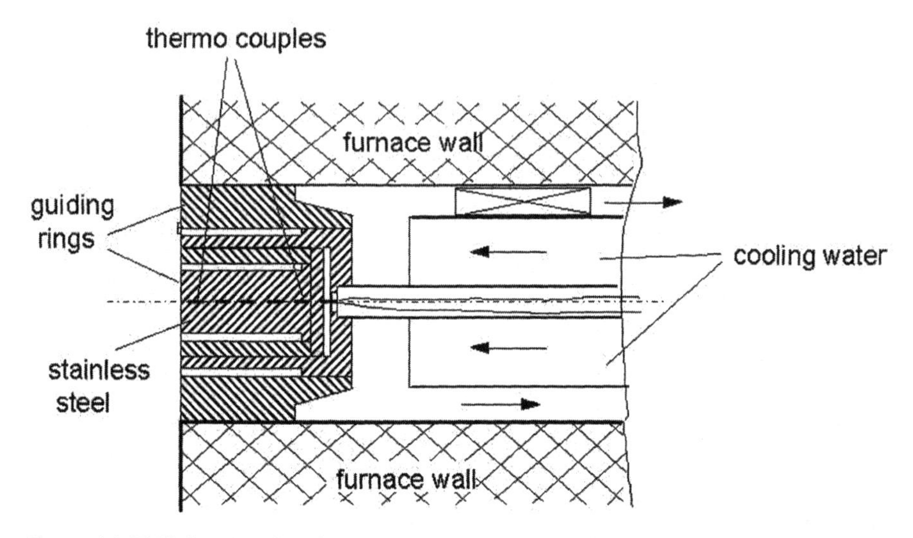

Figure 16. THF detector head.

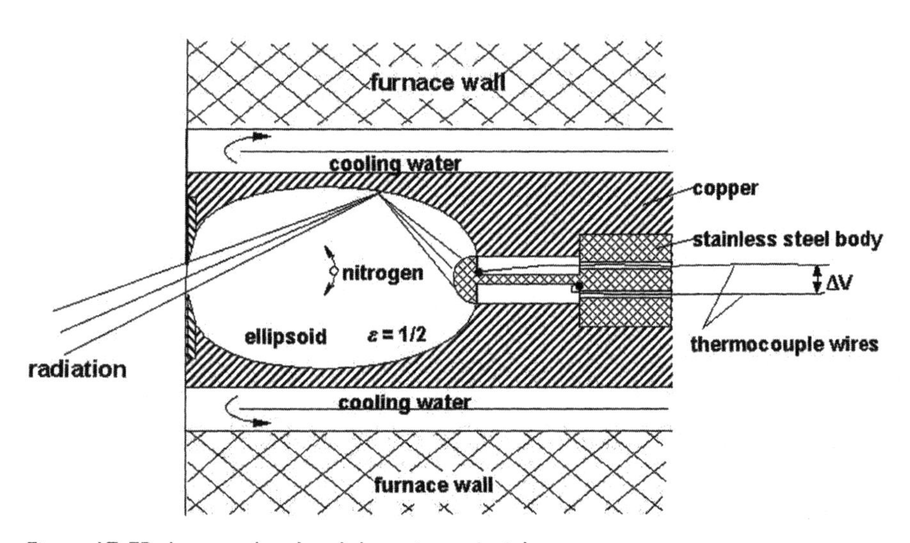

Figure 17. ER detector head and detection principle.

A sensor is situated at this point and conducts the heat received. The sensor features two resistors. Radiative heat flux is determined from the difference between the two resistances. The head of the ER is water cooled and purged by nitrogen to eliminate the measurement of convective heat transferred by the flow of hot flue gas and any influence of dirt.

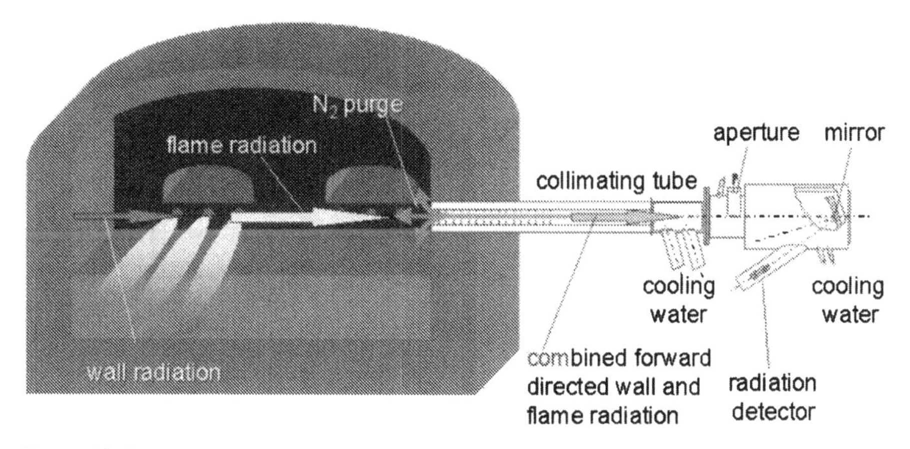

Figure 18. Detection principle of the NAR probe.

The narrow-angle radiometer (NAR) also measures total radiation, but because of the geometry of its collimating tube, all transverse radiation is absorbed by the cooling jacket of the probe and only horizontal radiation is transferred by a mirror to the sensor similar to that of the ER. The signal is weaker by several orders of size, though, and for this reason, the detector head must be cooled to maintain a constant temperature and the signal must be amplified. The probe is purged by nitrogen to eliminate dirt and to maintain constant conditions.[12]

The principle of radiation detection is shown in Fig. 18. The signals of both radiation detection devices (ER and NAR) are composed signals of both wall and flame radiation. These measurements at different locations along the furnace center line in the burner level were carried out to examine the influence of the different NO_x control techniques on heat transfer from the flame to the molten glass. In addition, the NAR signals can be compared easily with numerical simulated results.

Wavelength-Dependent Radiation Detection

With regard to the general objective of increased heat transfer from the flames to the load, spectral emissions measurements can be carried out. Especially for glass-melting furnaces, the radiation intensity at single wavelength is important.

Because a glass melt is a semi-transparent medium, radiation can be transferred into deeper layers of the glass tank. Therefore the spectral characteristics of the glass melt as well as of the flames must be investigated

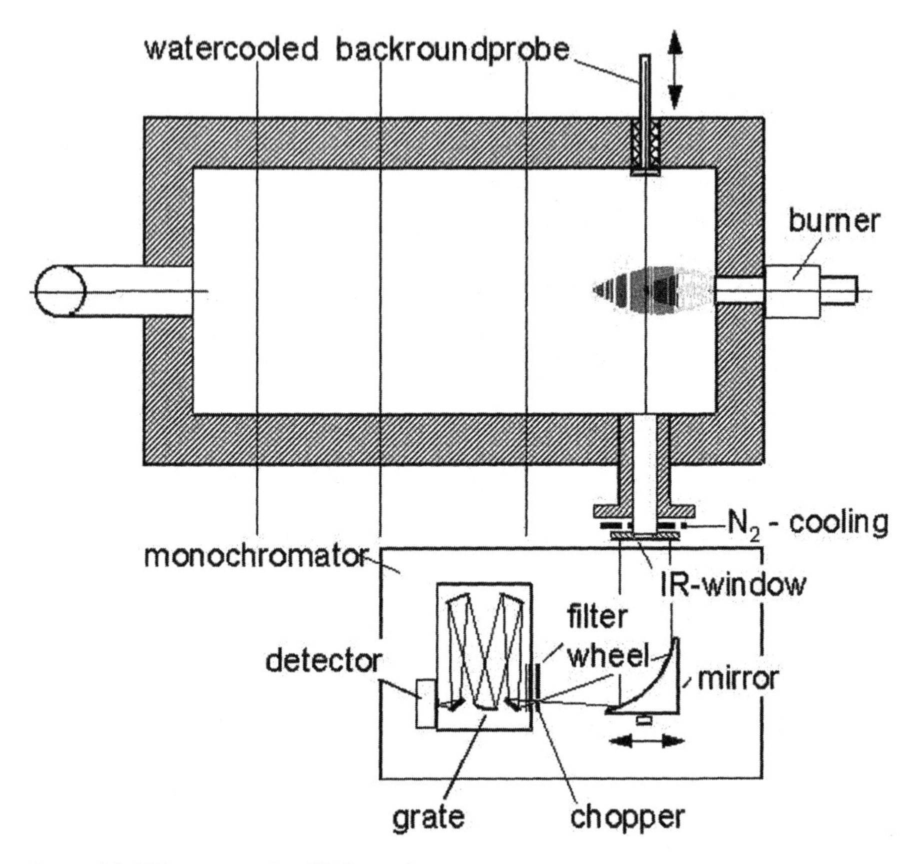

Figure 19. SPR setup at the GWI test furnace.

and adjusted with regard to each other. For this purpose a spectral radiometer (SPR) can be used (Fig. 19). The furnace wall radiation can be erased by using a cold target located at the opposite side of the furnace. Details of the measurement technique can be found in the literature.[13]

Summary

Different nonstandard measurement techniques for flow, mixture, combustion, and heat transfer detection were presented in this paper. Most of these techniques were developed for small flame laboratory style applications. During a longer time of operation the utilization area for this equipment was increased and the measurement system accuracy was improved. This leads to the conclusion that techniques in development today will be standard industrial methods within a few years.

The glass industry has realized strong improvements in the fields of combustion, energy savings, and pollutant emission reduction during the past decades. This success is based on investigations at glass tanks as well as in laboratories of industry, institutes, and universities where the measurement techniques described above have been applied.

Acknowledgments

The author would like to express his gratitude to industrial and research partners Hüttentechnische Vereinigung der Glasindustrie (HVG), Gaz de France-RD Division, DBI-Gas- u. Umwelttechnik (DBI-GUT), Netherlands Organisation for Applied Scientific Research (TNO-TPD), and the International Flame Research Foundation (IFRF) for their friendly cooperation.

References

1. J. Kompenhans, M. Raffel, and C. Willert, *Particle Image Velocimetry — A Practical Guide.* Springer, Berlin, 1998.
2. J. Westerweel, *Digital Particle Image Velocimetry: Theory and Application.* Delft University Press, 1993.
3. H. A. Becker, H. C. Hottel, and G. C. Wiliams, "On the Light-Scatter Technique for the Study of Turbulence and Mixing," *J. Fluid Mech.,* **30** [2] 259–284 (1967).
4. E. J. Shaughnessy and J. B. Morton, "Laser Light-Scattering Measurements of Particle Concentration in a Turbulent Jet," *J. Fluid Mech.,* **80** [1] 129–148 (1977).
5. R. E. Rosensweig, H. C. Hottel, and G. C. Williams, "Smoke-Scattered Light Measurements of Turbulent Concentration Fluctuations," *Chem. Eng. Sci.,* **15**, 111–129 (1961).
6. A. G. Gaydon and H. G. Wolfhard, *Flames, Their Structure, Radiation and Temperature.* Chapman & Hall Ltd., 1960.
7. H.-J. Voss and K.-W. Mergler "Visuelle Beobachtung nichtleuchtender Flammen in heißen Ofenräumen im UV-Bereich," Sonderdruck aus Glastechnische Berichte. 51. Jahrgang (1978), Verlag der Deutschen Glastechnischen Gesellschaft Frankfurt (Main).
8. A. Scherello, M. Flamme, and H. Kremer, "Optimization of Burner Systems for Glass Melting Furnaces with Regenerative Air Preheating"; presented at Advances in Fusion and Processing of Glass International Symposium, Ulm, 29–30 May 2000.
9. A. Scherello, M. Lorra, M. Flamme, and H. Kremer, "Neue Möglichkeiten zu NO_x-Emissionsminderung an Glasschmelzwannen," pp. 16–19 in *Proceedings of the 73rd Glastechnische Tagung.* 1999.
10. W. Van de Kamp, J. P. Smart, T. Nakamura, and M. E. Morgan, "NO_x Reduction and Heat Transfer Characteristics in Gas Fired Glass Furnaces." IFRF Research Report Doc. No. F 90/a/4. Ijmuiden, Netherlands, 1989.

11. W. Leuckel and A. Heilos, “Untersuchung zum Strahlungsverhalten rußhaltiger Kohlen-wasserstoffflammen,” *Gaswärme Intern.,* **43** [10] 1994.
12. N. Lallemant, A. Sayre, and R. Weber, “Evaluation of Emissivity Correlations of H_2O-CO_2-N_2/Air Mixtures and Coupling with Solution Methods of the Radiative Transfer Equation,” *Prog. Energy Combust. Sci.,* **22**, 543–574 (1996).
13. J. A. Wieringa, “Spectral Radiative Heat Transfer in Gas-Fired Furnaces,” Ph.D. thesis, Delft University of Technology, Delft, The Netherlands, 1992.

Alglass SUN: An Ultra-Low-NO_x Oxy Burner for Glass Furnaces with Adjustable Length and Heat Transfer Profile

Bertrand Leroux, Pascal Duperray, Patrick Recourt, and Rémi Tsiava
Air Liquide Claude-Delorme Research Center, Jouy-en-Josas, France

Nicolas Perrin
Air Liquide Chicago, Countryside, Illinois

George Todd
Air Liquide America Corporate, Tualatin, Oregon

A new oxygen burner technology called Alglass SUN (Separate Ultra-Low NO_x) has been developed by Air Liquide, which relies on a large separation of the fuel and oxidant streams and on the adjustable distribution of the oxidant in three different streams. Such a burner has been demonstrated to answer the dual needs of thermal profile optimization and minimization of NO_x emissions. Moreover, Alglass SUN soft flame seems particularly adapted to glass processes with its positive impact on glass quality and wall temperatures.

Introduction

The performance of a combustion system in a glass furnace is characterized mainly by two relevant parameters:

1. The control of heat transfer to the load, which is essential for glass quality and process performance.
2. Pollutant emission levels (NO_x, dust, etc.), which must meet more and more stringent regulations.

The effect of some burner and furnace parameters on the heat transfer profile along the flame axis has already been put into evidence by numerous studies.[1–3] Flame shape and burner momentum appear as the two most relevant parameters. A long, high-momentum flame is recommended to maximize heat transfer away from the burner; moreover, such operating conditions enable the reduction of the maximum local flame temperature along the furnace axis, and the maximum furnace wall temperature, A short, low-momentum flame is recommended to maximize heat transfer close to the burner.

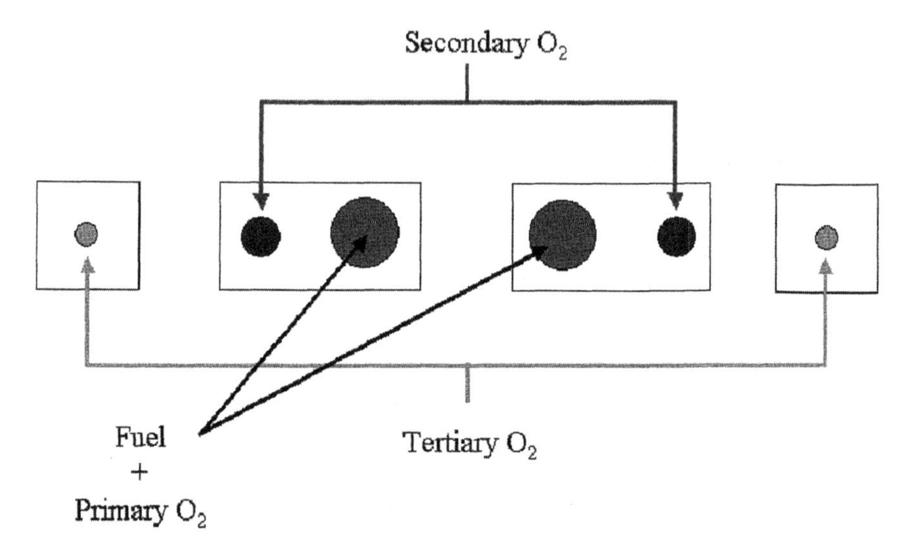

Figure 1. Alglass SUN burner geometry.

Previous flame characteristics (long and high-momentum on the one hand; short and low-momentum on the other hand) cannot be reached with standard oxy burners (for example, pipe-in-pipe burners), even with double impulse injections for oxidant. In these geometries where oxidant flow is located near fuel injection, increasing the flame momentum causes an acceleration of the mixture, a reduction in equivalent burner diameter, and consequently a decrease of flame length.

The Alglass SUN burner has been designed to meet desired flame specifications. This is a double impulse system (see Fig. 1) with an oxidant flow located at a certain distance of the fuel injection (secondary O_2) and a second one positioned at a larger distance (tertiary O_2). Because the diameter of secondary O_2 jet is more important than the tertiary O_2 one, changing the repartition between these two injections enables modification of the flame momentum and length according to required characteristics. Indeed, when increasing the oxidant repartition through the smaller and more distant injection (tertiary O_2), flame momentum becomes higher and reactant jets entrain an important flow rate of combustion products: combustion is delayed and takes place over a longer and larger volume. Note also that a third oxidant flow (primary O_2) adjoining the fuel injection has been added to initiate combustion and guarantee flame stability.

Another requirement for this new oxy burner was a significant reduction in NO_x emissions compared with standard burner geometries. Some data used for such a development have been obtained during the Oxyflam program.[4] The analysis of these experimental results has shown that NO_x emissions were controlled primarily by the degree of dilution of incoming reactants by combustion products contained in the furnace atmosphere. Consequently, to obtain low peak flame temperatures and low thermal NO_x formation rates, the main parameter to take into account is the ratio of the oxygen-to-fuel distance over the oxygen nozzle diameter.

The Alglass SUN design answers previous specifications and is optimal in reaching ultra-low NO_x levels while maintaining high transfer to the load and avoiding volatilization issues. The main part of oxidant (between 50 and 75%) is injected toward the tertiary flows at a large distance of fuel injection while primary O_2 flow rate is maintained low enough to avoid too-high local temperatures.

Two other specifications regarding this new burner should also be noted:

1. Compared to traditional burners based on a single ceramic block, this burner consists of two types of small ceramic blocks: one for the tertiary oxygen lances and one for the fuel, primary, and secondary oxygen injectors. A ceramic block height of 150 mm is sufficient to accommodate injectors for burner firing rates up to 5 MW. Such a geometry may be easily installed in existing air firing furnaces for oxy boosting, in particular in the case of underport geometries.
2. Alglass SUN may be used either with natural gas or heavy fuel oil by changing only the fuel lance.

Procedure

The benefits provided by the Alglass SUN burner were first underscored at a nominal firing rate of 2 MW in an Air Liquide pilot furnace (see Fig. 2). This ceramic fiber–lined furnace has a low thermal inertia and can operate at a wall temperature up to 1600°C. The combustion chamber is 6.0 m long, and has a rectangular cross section of 1.5 × 2.0 m. These internal dimensions are large enough to minimize flame confinement effects and allow investigation of the interaction between the flame aerodynamics and the furnace recirculation zones. Based on constant velocity scaling rules, exact aerodynamic similarity is obtained between a 2.0 MW flame in the CRCD furnace and a 5.0 MW flame in a furnace with a length of 9.5 m and a cross section of 2.37 × 3.16 m.

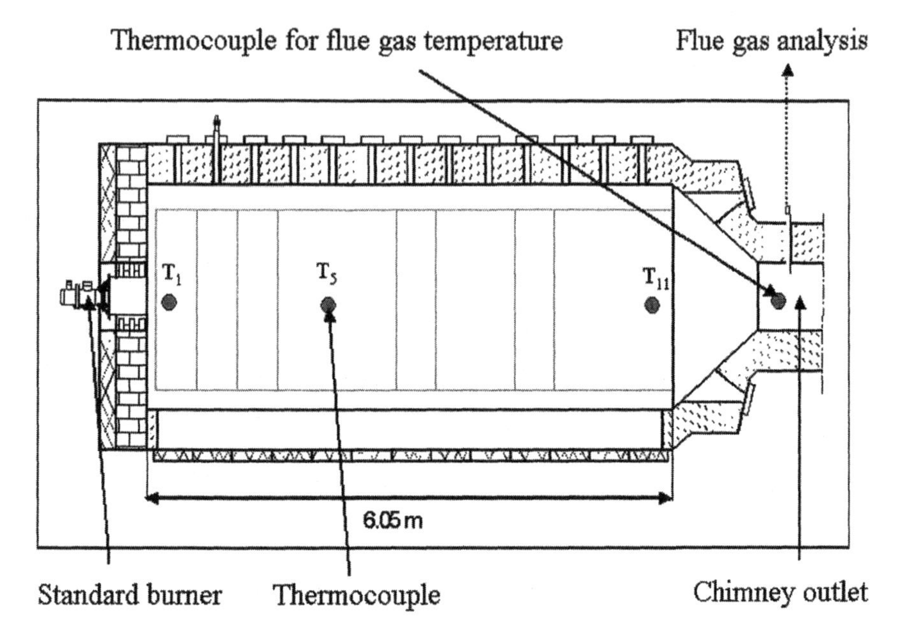

Figure 2. Air Liquide CRCD pilot furnace.

Heat extraction and furnace outlet temperature could be varied by insulating the sides of the water-cooled furnace floor and by covering part of the water-cooled panels with a radiative screen consisting of silicon carbide panels. The furnace roof axial temperature profile was measured by 11 Type S thermocouples. Average furnace temperature is calculated with data given by these 11 thermocouples. The heat extraction profile was obtained by calorimetric measurements over the 13 water-cooled floor panels. A set of analyzers and a suction pyrometer measured the flue gas composition (O_2, CO, CO_2, and NO_x) and temperature.

Video cameras were positioned in the furnace side and above the furnace horizontal exhaust section. Image analysis was achieved; in particular, fluctuation images were calculated in order to put into evidence luminosity fluctuation and quantify flame contour. In this case, fluctuation image was defined as the difference between current and average images.

An electric damper in the chimney duct allowed regulation of furnace pressure and minimization of air leakage into the furnace. (Further details on this furnace and its diagnostics can be found in Ref. 5.) The level of

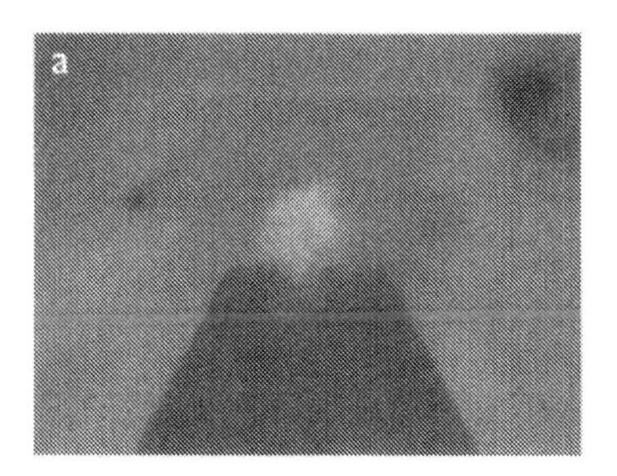

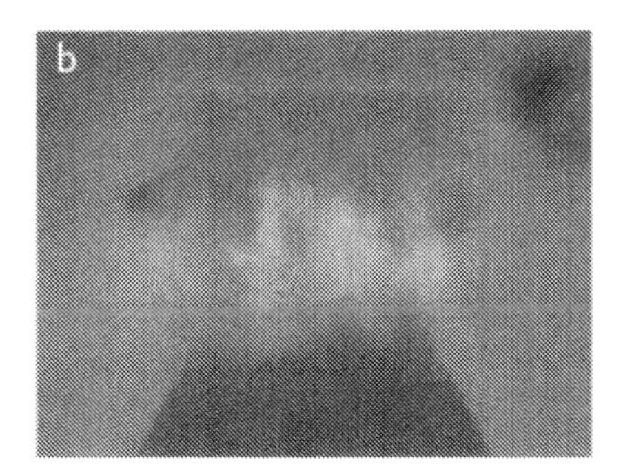

Figure 3. Alglass SUN live flame (face view; power = 2 MW; natural gas). (a) Tertiary oxygen ratio = 50%. (b) Tertiary oxygen ratio = 75%.

nitrogen concentration in the furnace dry flue gas was taken as the complement of the (CO_2+O_2) concentrations to 100%. Care was taken to record time-averaged data for stable input and output parameters.

Experimental characterization of the burner prototype involved parametric measurements of furnace heat transfer profile, flue gas temperature, and NO_x concentration. A total of 400 firing modes were measured using successively natural gas or heavy fuel oil, and different atomization gas types (air, O_2, steam, natural gas in heavy fuel oil case).

Results

Influence of Tertiary Oxygen Ratio on Flame Structure and Axial Heat Transfer Profile

The influence of tertiary oxygen ratio on flame structure is shown in Fig. 3 for the natural gas case and in Fig. 4 for the heavy fuel oil case. Instantaneous images are shown according to face view for two oxidant repartitions. The first corresponds to a tertiary oxygen flow rate equal to 50% of the total oxidant flow rate; the second, to a tertiary oxygen flow rate equal to 75% of the total oxygen flow rate. In both cases, burner power is equal to 2 MW and furnace temperature is about 1500°C.

For the lowest tertiary oxidant flow rate (50%), the flame has a lower volume (about 3 m long and 1.5 m wide), whereas for the highest tertiary oxidant flow rate (75%), flame volume becomes very large because the flame is about 2 m wide and 4.5 m long.

Flame volume evolution as a function of tertiary oxygen ratio has an influence on axial heat transfer. Figure 5 shows that low combustion stag-

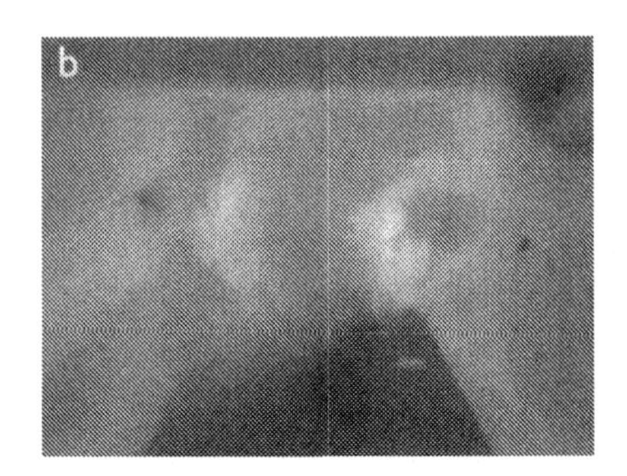

Figure 4. Alglass SUN live flame (face view; power = 2MW; heavy fuel oil; air atomization).

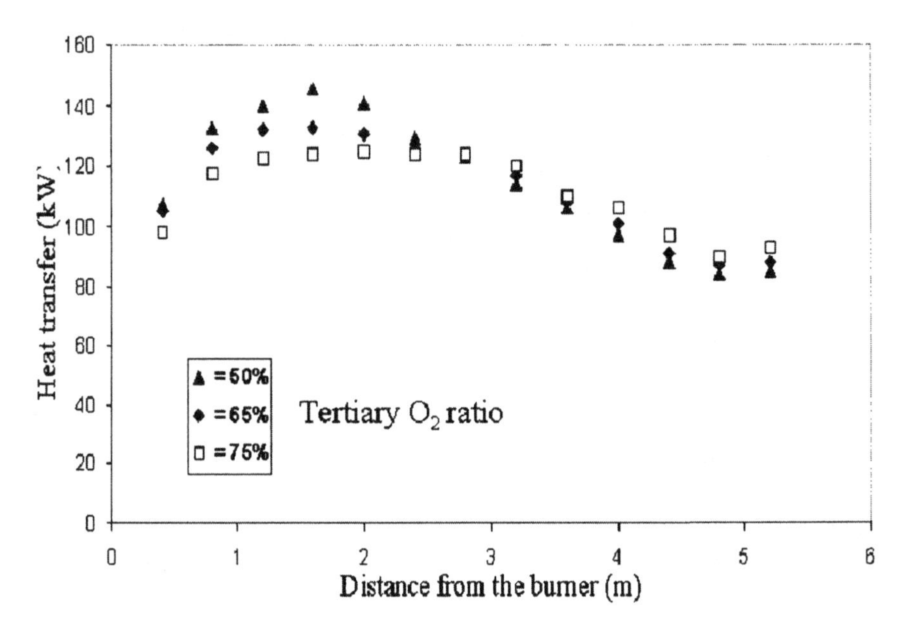

Figure 5. Effect of tertiary oxygen ratio on axial heat transfer (power = 2 MW; fuel oil; O_2 atomization).

ing (50% oxygen in tertiary flow) enables to maximize heat transfer close to the burner whereas high staging (75% tertiary oxygen) allows maximizing heat transfer away from the burner. In the latter case, thermal profile is much more homogeneous because the maximum difference on thermal flux is about 30 kW. Consequently, the design of this new burner answers one of

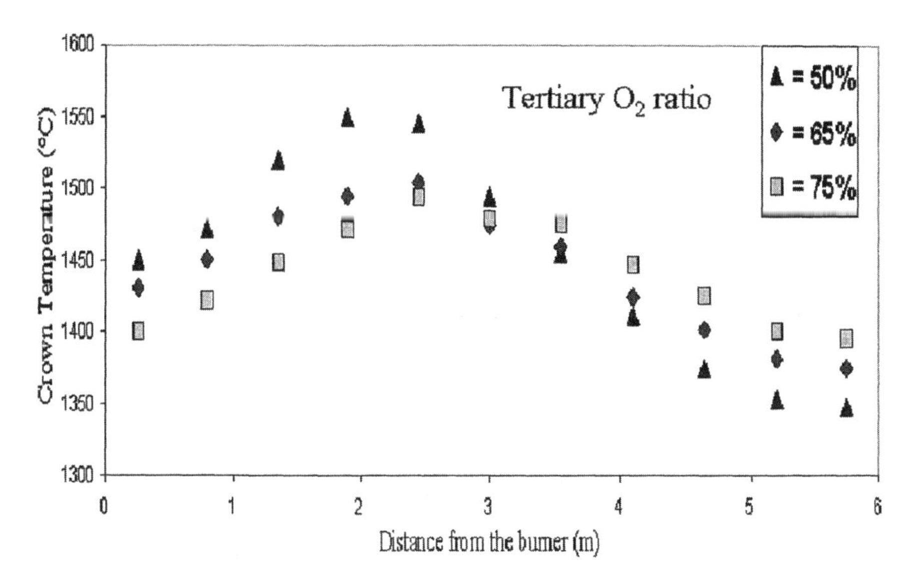

Figure 6. Effect of tertiary O_2 ratio on furnace axial temperature (power = 2 MW; natural gas).

the previously defined purposes: a large flexibility for axial heat transfer distribution. It must be pointed out that the value of the tertiary oxygen ratio yielding the highest heat extraction may be furnace dependent, and in particular on the position of the flue gas duct compared to the burner position. It is expected that the ability to fine-tune the heat transfer profile along the flame axis should translate into improvement in product quality in many furnace types.

Flame volume evolution has also an effect on furnace local temperatures. Figure 6 displays the evolution of axial crown temperature as a function of tertiary oxygen flow rate. It shows that increasing the tertiary oxygen ratio from 50 to 75% increases the furnace back-end temperature by 55°C while decreasing the furnace front-end and maximum roof temperatures by 50°C. It may be shown that increasing the tertiary oxygen ratio from 50 to 75% leads indeed to an increase of the total burner momentum; as observed from experience and internal studies, the increase in burner momentum has a beneficial impact on the furnace temperature uniformity.

NO_x *Emissions*

Figure 7 shows the effect of tertiary oxygen ratio on NO_x emissions when operating with natural gas. For all data shown in Fig. 7, the nitrogen con-

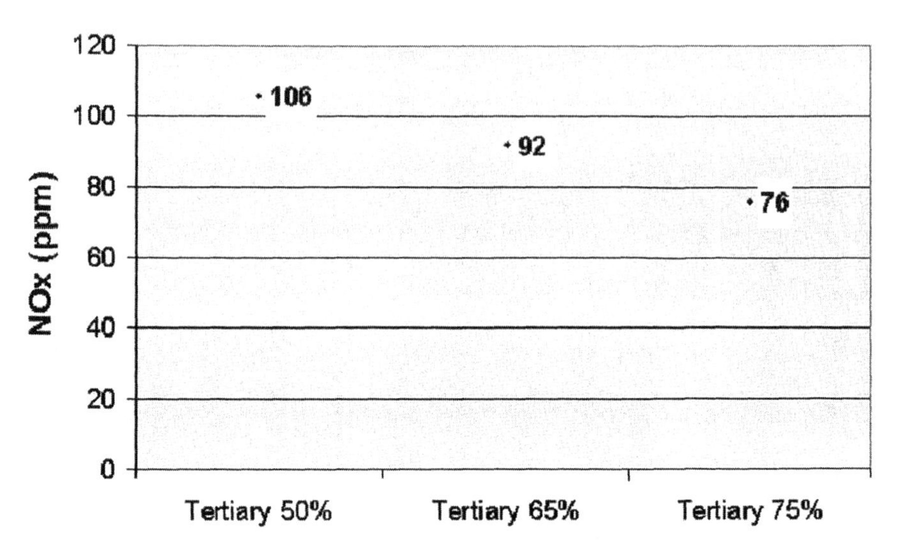

Figure 7. Influence of tertiary oxygen ratio on NO_x emissions (power = 2 MW; natural gas).

centration in the dry flue gas was around 7.0%. The results show that for an average furnace temperature equal to 1510°C, the NO_x emissions range from 76 to 106 ppm. Because pure oxygen combustion leads to a dry flue gas volume reduction by a factor of about 9.5, the NO_x emissions indicated above are equivalent to NO_x concentrations of 11–16 ppm in the combustion products from air combustion.

Throughout the experiments, it was noted that tertiary oxygen ratios above 75% led to large pressure fluctuations and increased air aspiration into the furnace. Under extreme conditions, a flame with a 95% tertiary oxygen ratio exhibited important changes in the flame reaction zone, with two main, luminous combustion zones located around the tertiary oxygen jets. From the instantaneous furnace pressure data, the standard deviation of the pressure measurements was calculated. The furnace pressure standard deviation is about constant for a tertiary staging ratio from 0 to 75%, and increases sharply in the range from 75 to 95%. This effect was observed experimentally from the increase in nitrogen content in the flue gas. Under very high tertiary oxygen staging conditions, closing the chimney damper and increasing the mean furnace pressure to 2.5 mm H_2O could not prevent the increase in nitrogen concentration in the flue gas. Thus, although a high staging ratio is beneficial to minimize NO_x emissions, it appears preferable

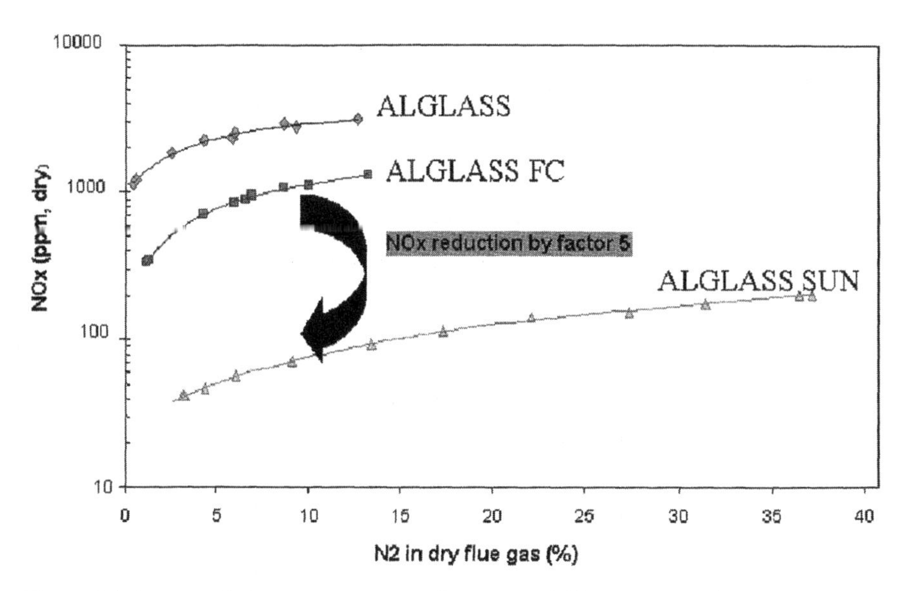

Figure 8. Influence of nitrogen content in flue gas on NO_x emissions (comparison with other Air Liquide oxy burners; power = 2 MW; natural gas).

to not exceed a secondary oxygen ratio of 75% in order to keep pressure fluctuations and air aspiration to a minimum. The standard deviation of the instantaneous furnace pressure can be used as a useful indicator to optimize the burner oxygen distribution.

The influence of nitrogen concentration in flue gas on NO_x emissions has also been studied for natural gas and is displayed in Fig. 8. Average furnace temperature is about 1510°C and tertiary oxygen ratio is equal to 75%. In the present experiments, the nitrogen concentration was increased by opening the chimney damper, which reduced the furnace pressure and increased the air aspiration into the furnace. The furnace average temperature was maintained constant by increasing the firing rate to compensate for the aspiration of cold air. The oxygen flow rate was also adjusted to maintain 3% O_2 in the dry flue gas.

Figure 8 compares NO_x levels obtained in the same firing conditions with two other Air Liquide technologies: the Alglass burner, which was the first self cooled oxy burner developed by Air Liquide, and the Alglass FC burner, which is a standard staged oxy combustion technology. When comparing with a low NO_x oxy burner such as Alglass FC, the Alglass SUN burner enables the reduction of NO_x levels by a factor of five.

Conclusions

The performance of the Alglass SUN burner was characterized by parametric measurements of furnace heat transfer profile, flue gas temperature, and NO_x concentration for a total of 400 firing modes. The recommended oxygen distribution oxygen ratio is a tertiary oxygen flow rate varying between 50 and 75% of the total oxygen amount, depending on the required flame heat release and heat transfer profile, with the remaining oxygen flow introduced through the primary and secondary oxygen injectors.

It was shown that by varying the tertiary oxygen ratio between 50 and 75%, the furnace heat transfer profile along the flame axis could be adjusted so as to reduce the maximum crown temperature by 50°C and simultaneously increase the furnace back end temperature by 50°C. Thus, if burner thermal input is maintained as a constant, the difference between the maximum crown and the minimum back end temperatures could be varied from 190°C to a minimum of 90°C. The variations in heat transfer profile showed a minimum impact on ultra-low-NO_x performance and overall furnace heat transfer efficiency. When compared to standard burners with fixed characteristics, the ability to optimize the flame properties and heat transfer profile to each industrial furnace geometry is expected to have a positive impact on product quality, refractory wear, and fuel efficiency.

When using natural gas as fuel, this burner demonstrated NO_x emissions between 76 and 106 ppm (equivalent to 4 to 6 mg/MJ), at 3% O_2 and 5% N_2 in dry flue gas, for average furnace temperatures equal to 1510°C. For a nitrogen concentration of 50% in dry flue gas, the NO_x emissions would be about 400 ppm. This ability to maintain ultra-low-NO_x emissions, even under high nitrogen concentrations in the furnace gases, allows implementation of this technology in industrial furnaces fired with air and oxygen burners. Furthermore, because the Alglass SUN burner consists of two types of small ceramic blocks (150 mm high for burner firing rates up to 5.0 MW), this burner is very easy to implement in existing geometries.

A configuration similar to the Alglass SUN burner has been tested in a large glass container furnace during the repair of lateral regenerators and the switch to full oxy firing conditions. These results cannot be detailed here for confidentiality reasons. However, the conclusions of this test have been very promising. In particular, the Alglass SUN soft flame seems particularly adapted to glass process with a positive impact on glass quality and wall temperatures.

Acknowledgments

The authors wish to acknowledge J. Dugué for the development of the highly separate jets concept and the contributions of B. Grand and L. Sylvestre for the furnace preparation. The authors also wish to thank C. Imbernon for assistance in post-processing of the furnace data.

References

1. H. C. Hottel and A. F. Sarofim, *Radiative Transfer.* McGraw-Hill, New York, 1967.
2. J. M. Beér and N. A. Chigier, *Combustion Aerodynamics.* Applied Science Publishers, 1972.
3. S. Michelfelder and T. M. Lowes, "Report on the Mathematical Modeling M-2 Trials." IFRF Doc. F36/a/4, August 1974.
4. N. Lallemant et al., "Flame Structure, Heat Transfer and Pollutant Emissions Characteristics of Oxy-Natural Gas Flames in the 0.7-1 MW Thermal Input Range." *J. Inst. Energy,* **73** [September] 169–182 (2000).
5. J. Dugué, W. Von Drasek, J. M. Samaniego, O. Charon, and T. Oguro, "Advanced Combustion Facilities and Diagnostics"; presented at the American/Japanese Flame Research Committees International Symposium, Maui, Hawaii, 11–15 October 1998.

Glass Furnace Life Extension Using Convective Glass Melting

Neil Simpson and Dick Marshall
BOC Glass

Tom Barrow
Pilkington

The float glass industry has experienced global growth, which has consistently resulted in the construction of approximately three new furnaces per year. Currently, several glass companies operate more than 20 float furnaces each, which places considerable demand on their capital for furnace repairs and growth. Therefore, extending the life of these furnaces while maintaining production efficiency and thus delaying capital expenditures is becoming a priority for all the major float glass manufacturers. BOC's convective glass melting (CGM) technology offers a unique method to extend the life of a furnace for several years while maintaining or increasing yield. This paper will compare conventional zero boost technology with the application of CGM to a float furnace. A case study will review a float glass melter with significant regenerator failure, which required CGM to supply up to 30% of the total melting energy (hybrid melter) for a one-year life extension. The paper will review the technical and financial considerations of the CGM application and its relevance to other glass market segments.

Conventional Zero Port Boost Technology

The use of oxy-fuel burners to boost glass furnaces is a proven technology, specifically in float furnaces that normally incorporate a 2–4 m distance between the charge end wall and the first port jamb block. Figure 1 illustrates the application of this boosting technology.

The 50 furnaces worldwide that have implemented conventional zero port oxy-fuel technology over the last five years have provided the industry with significant understanding of oxy-fuel boosting's capabilities and limitations. The addition of 10–15% more energy into the furnace in the critical melting area can result in an increase in pull rate, quality, and furnace life.

The results of oxy-fuel boosting have been mixed at times, primarily because of a failure of the parties involved to establish a defined objective prior to installing the boost systems. The four most common reasons for installing oxy-fuel boost in glass furnaces are to increase pull rate, improve product quality, extend furnace life due to regenerator checker failure, and extend furnace life by preventing regenerator checker failure.

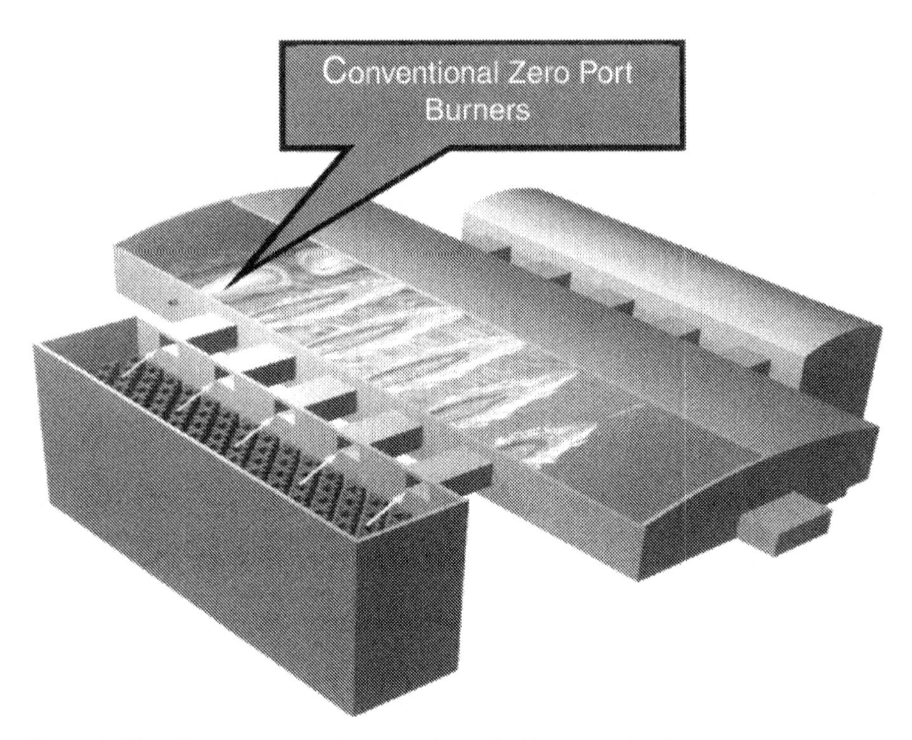

Figure 1. Glass furnace using conventional oxy-fuel boost technology.

Each of these objectives requires a different approach for adapting oxy-fuel boosting to a regenerative furnace. For example, when increasing the pull rate of the furnace, the batch entering the charge end of the melter absorbs most of the additional energy supplied by the oxy-fuel burners, and the air-fuel combustion system remains essentially unchanged. However, when extending the life of the furnace due to regenerator checker failure, the objective is to minimize the overall products of combustion within the system. To be most effective, the applied oxy-fuel energy must be removed on a total or overall basis from the air/fuel port energy, because there are very limited means of reducing the air flow on a given port. This may require a change in the furnace temperature operating schedule.

Convective Glass Melting Technology

The zero port breastwall area limits the amount of energy that can be introduced from conventional boost technology (Fig. 2). If this limited production, quality, or life extension is sufficient to meet the glass manufacturer's

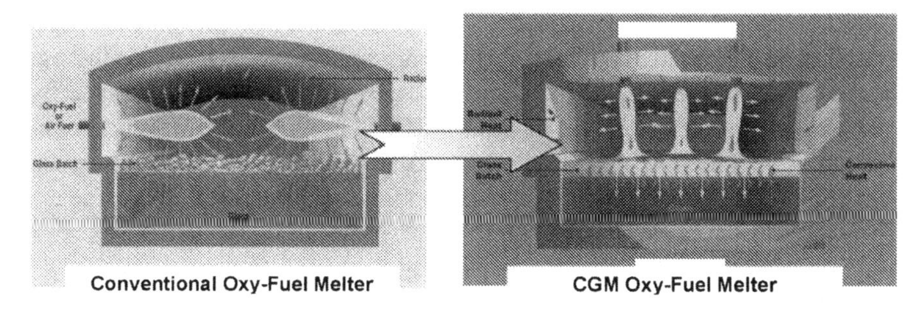

Figure 2.

objectives, then conventional (breastwall mounted) combustion is the preferred technology.

BOC's convective glass melting (CGM) technology increases in the rate of heat transfer to the batch and thus increases the rate of melting. This benefit removes many of these limitations of conventional zero port boost and can provide increased performance improvements to many glass melters.

BOC, in cooperation with Owens-Corning's composites fiberglass division, introduced the first successful, vertically oriented, crown-mounted, oxy-fuel burners on a commercial-size glass melting furnace in 1995. This three-month test concluded that CGM could substantially increase the rate of batch melting over a side-fired air-fuel or oxy-fuel furnace. Because these burners are located in the crown, the space limitations are overcome and the number of burners needed to meet the project objectives can be installed. In multiple cases, BOC has fired over 50% of the total energy of the furnace through the CGM burners, with less than 50% remaining on the air-fuel burners.

Since the first commercial installation of CGM in 1996, it has been installed in over 20 operating glass melters. They include:

- Melters for the float, container, fiberglass, and tableware markets.
- Oxy-gas and oxy-oil burners.
- Air-fuel and oxy-fuel furnaces.
- Asia, North America, and Europe.

BOC filed patents to protect the CGM process as the technology continued to be developed. The first patent (U.S. 6 237 369) was filed jointly with Owens Corning on 10 February 10 1997 and was issued on 29 May 29 2001. The second patent (U.S. 6 422 041) was filed on 16 August 1999 and issued on 23 July 2002. This patent discloses the use of CGM burners positioned over the raw batch materials in an air-fired furnace. It details the use of the CGM burners to initially boost a three-port side-fired furnace by blocking off the first port and installing CGM burners. A CGM hybrid configuration was then achieved by additionally blocking off the second port, installing CGM burners, and continuing to fire the third port with air-fuel.

Case Study: Float Furnace Life Extension Using CGM

The float segment of the glass industry has continued to show global growth of about 2000 tpd of new product demand annually. As the major float glass suppliers install new furnaces, the number of furnace repairs per year also increases. This has placed a significant demand on capital to concurrently maintain growth and current business activity. Also, because many of the major float suppliers now have in excess of 25 furnaces, the scheduling of these repairs often conflicts with human resources as well as capital. Therefore, if the life of a furnace can be extended, the capital allocated for that repair can be delayed, which helps provide a significant cash flow benefit. CGM has proven to be an effective way of providing continued operation of a float glass melter even when the checkers are significantly damaged.

In 1999, Pilkington's Laurinburg float furnace was operating as a seven-port regenerative furnace with conventional oxy-fuel boost burners firing at 20 MMBtu/h (Fig. 3). The furnace was operating at 700 tpd with a need for increased product. In addition, after 13 years of operation, the furnace had serious regenerator checker damage, which was threatening an early repair. As a result of product demand and lack of materials in inventory for the repair, it was essential to maintain furnace operation for a minimum of one year. After reviewing several technologies, including port oxygen lancing and hot checker repair, CGM was chosen as the life extension technology.

The installation of the CGM on the glass furnace (Fig. 4) consisted of the following steps:

1. A seal coating of silica castable was installed over the silica crown. The silica castable seal coat is important in crowns to protect it against the problems associated with the higher alkali concentrations that exist in oxy-fuel furnaces. Crown wear around existing ratholes will accelerate without this seal coat. It is recognized that this wear is the general result of the oxy-fuel atmosphere and not the CGM.

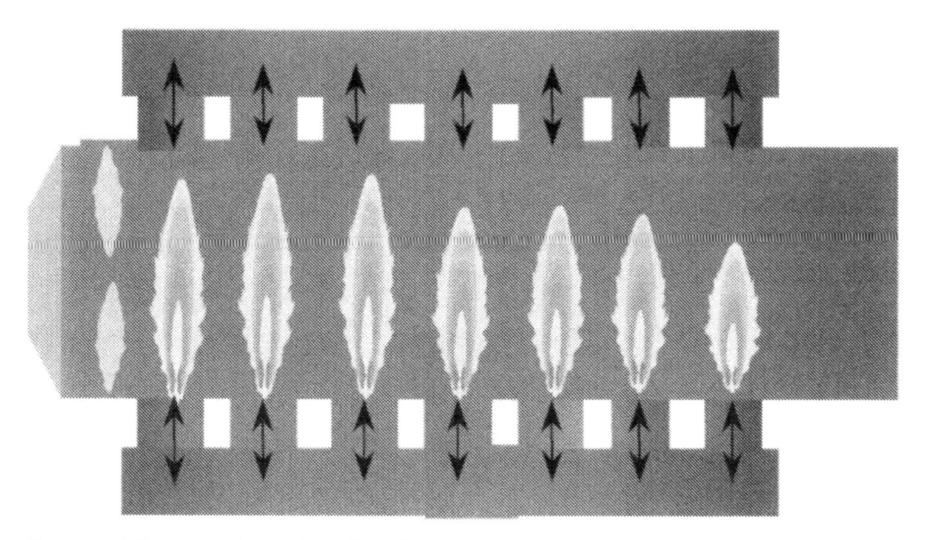

Figure 3. Pilkington's Laurinburg float furnace.

Figure 4. Typical CGM furnace installation.

2. A crown access platform was installed over the section of the crown where the CGM burners were to be installed.
3. Six holes were drilled in the silica crown while the furnace was in production.
4. The combustion piping system was installed, which included a multizone flow control skid.
5. Damper gates to isolate the first port were purchased and installed.

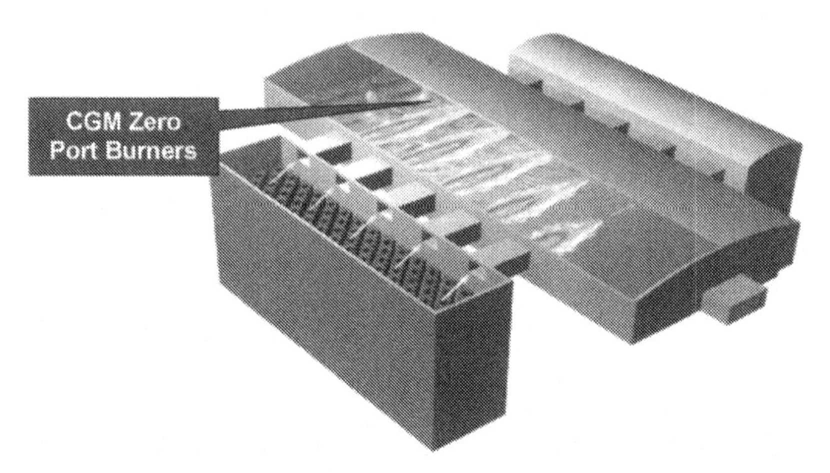

Figure 5. Pilkington's Laurinburg float furnace with CGM zero port operation. (Shown as a five-port furnace for simplicity.)

The only negative effect on production during this time was an incremental increase in stones resulting from the drilling process.

On 1 November 1999, two CGM burners were placed in the zero port area of the crown and fired at the same total rate of 20 MMBtu/h as the sidewall burners, which were simultaneously removed (Fig. 5). This action was intended to give plant operations time to observe the CGM operation prior to expanding CGM to the hybrid configuration. The conversion from sidewall to CGM operation was accomplished while in normal production with no adverse effects on production. The primary observations were a noticeable increase in glazing of the batch as it exited the doghouse into the melter and the receding of the batch line by approximately one-third of the width of a port.

In December 1999, Port 1 was blocked off and four additional CGM burners were installed in the approximate positions shown in Fig. 6. This conversion was conducted with the furnace in operation with little observable effect on the production capacity or select rate. The CGM burners were initially adjusted to fire at 77% of the air-fuel firing level of Port 1. This maintained the furnace in approximately the same operation (including batch line, pull rate, and quality) as it had been prior to the installation of the additional four CGM burners.

While the initial objective of installing CGM was to increase the pull rate above that prior to its installation, the continued deterioration of the

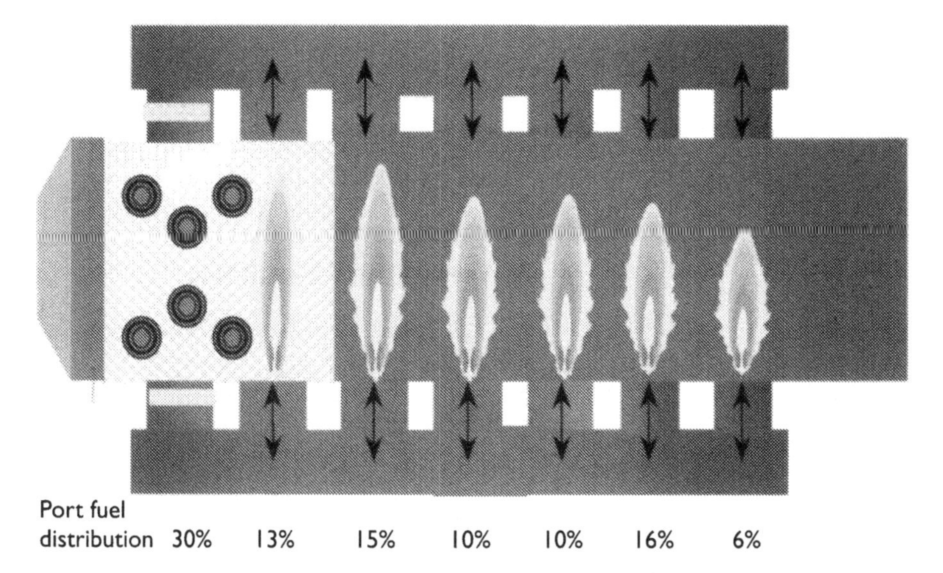

Figure 6. Pilkington's Laurinburg float furnace. Fuel distribution with CGM.

furnace regenerators placed an increased demand on CGM to simply maintain the existing production. The CGM firing rate was increased to a maximum of 65 400 scfh or approximately 30% of the furnace firing capacity. Initial consideration was given to extending the CGM to Port 2, but the availability of oxygen supply in the local region prohibited this action. Figure 7 is a photograph of two of the CGM burners in operation filmed during a reversal of the air-fuel combustion system.

The furnace operated in this configuration until its repair in December 2000, a total of 14 months. Because of the continual deterioration of the regenerators, it

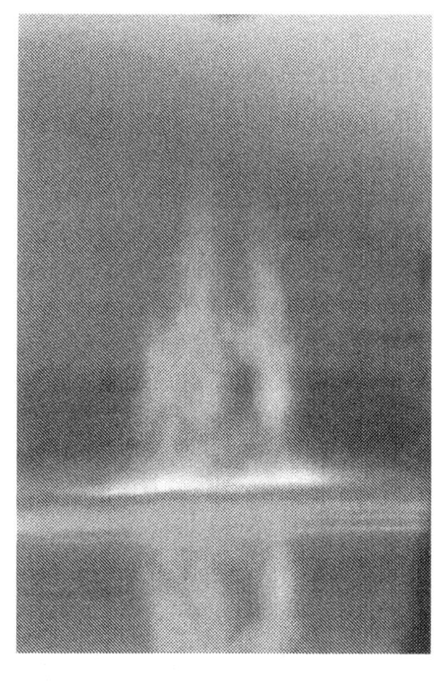

Figure 7. Pilkington Laurinburg furnace CGM flames in operation.

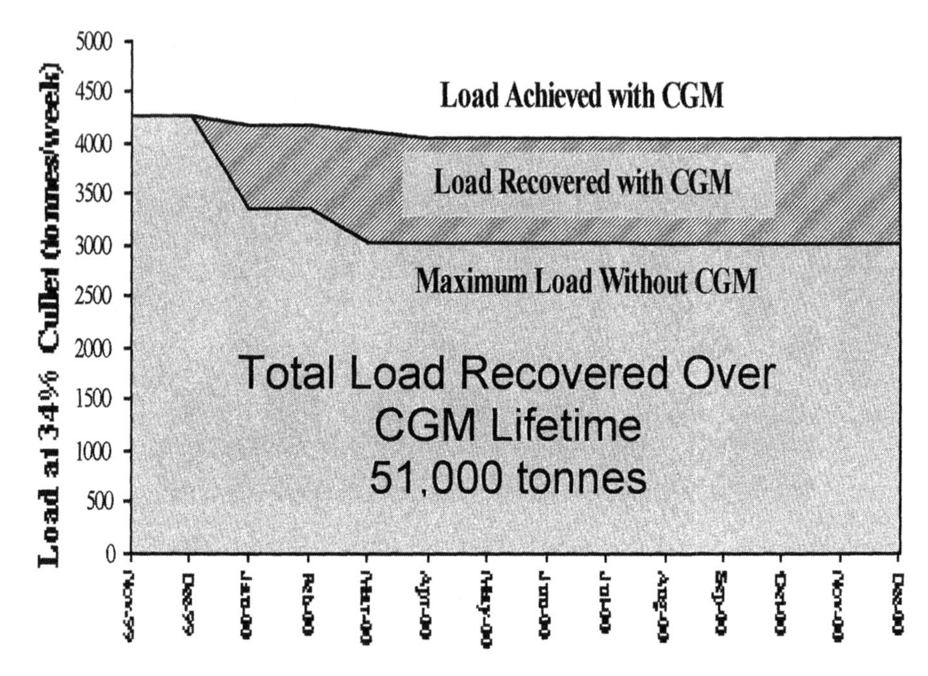

Figure 8. Pilkington Laurinburg CGM benefit.

was difficult to assess the absolute value of CGM from daily observations. Therefore, Pilkington engaged in a thorough analysis based on established operational practices and CFD modeling.

Figure 8 illustrates the benefits of implementing CGM on this furnace as a function of recovered tons melted. The primary conclusions of this study were as follows:

- CGM provided 51 000 tonnes more product than could have been achieved otherwise, which is approximately a 25% increase in melting capability.
- The furnace melting capacity without CGM may not have been sufficient to maintain the ribbon; hence production would have halted prior to the scheduled repair date.

General furnace conditions after 14 months of firing CGM were no worse than expected for a furnace of its age. The silica crown did deteriorate slightly, as expected, under oxy-fuel combustion and there was some incremental wear directly around the burner blocks. This increased wear is believed to be attributed to alkali condensation around the burner tube, which is cooled by the oxygen gas flow through the tube.

While this application was for life extension, it was in reality a cross between an oxy-fuel and an air-fuel furnace, or a hybrid configuration. The concept of melting with oxy-fuel CGM and refining with conventional air-fuel combustion is worthy of consideration in future furnace designs. This hybrid configuration has the following advantages:

- Lower conversion capital cost due to considerably less fused cast alumina.
- Lower operating cost due to reduction in oxygen consumption.
- Lower NO_x formation due to reduced air-fuel combustion.
- Better foam control using conventional technologies.
- Heat recovery of the oxygen products of combustion by preheating of the combustion air.

Financial Review

The 51 000 tonnes of product over a 14-month period furnace life extension/load recovery at Pilkington Laurinburg represents a significant return on investment with a relatively short payback period for a glass manufacturer. A conservative benefit of $1–2 million can be attributed to delaying a furnace repair by one year, based only on the cost of capital.

If the increased melting capacity is not required, an improvement in product quality or yield can provide substantial financial rewards. The noticeable receding of the batch line at Laurinburg using the CGM system can be translated into higher yields for the manufacturer. A 1% yield improvement will result in additional 5–7 tpd of product, which translates into a profit of approximately $500 000 per year.

Float glass manufacturers realize that oxygen boost is easily justified when they can use the increased production or improve yields by a reduction in defects. However, they do not want to increase their operating costs when the need for increased production or quality doesn't exist. This has led to the choice of liquid oxygen as the preferred method of supply chosen by most glass manufacturers. A primary attraction is the short-term supply commitment that matches the need of the user. However, the unit cost of oxygen delivered through this mode is substantially higher than that delivered through an on-site supply.

While an on-site supply solution may require a much longer term of commitment, the product cost is substantially lower than liquid product. One approach is to install an on-site supply toward the end of a furnace

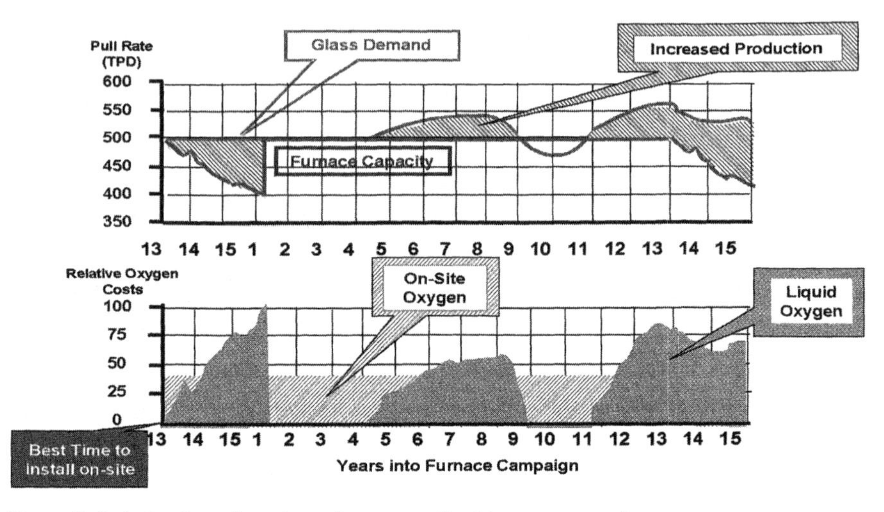

Figure 9. Relative benefits of on-site versus liquid oxygen supply.

campaign and to opportunistically use the lower cost oxygen supply for the remainder of the agreement during the new campaign. While the increase in product or quality may not typically be required until several months after the repair, oxygen boost could be used to reduce overall fuel consumption in the interim. Additionally, lower cost on-site oxygen provides the glass manufacturer tremendous flexibility in ramping production as needed, even in the earlier years of the new campaign or during periods of short-term product demand.

Figure 9 illustrates an assumed furnace capacity and product demand cycle through the life of a typical float furnace. The capacity of a float furnace (assumed to be constant at 500 tpd in Fig. 9) often deteriorates during the last few years of its life, providing the opportunity to recover that lost capacity using conventional oxy-fuel or CGM oxy-fuel boost. Also, during the 15-year life of new float furnaces, product or quality demands can exceed the capacity of the furnace. Liquid oxygen is an effective way of meeting this increased demand and because of its short-term supply agreement, it can be removed when the product demand is no longer there. An on-site oxygen supply system lowers the unit cost of oxygen supply, but the longer term agreement requires continual consumption of the gas produced. This does increase furnace operating costs when the additional production is not required, however it is offset by a reduction in overall melting energy (demonstrated to be 30% of the applied boost energy).

Several parameters need to be considered in the decision making process, and the answer is very much project-specific. However, the results of our general analysis suggest that an on-site approach should be considered prior to implementing an oxy-fuel boost operation.

Future Applications of CGM

The application of CGM to float glass furnaces for life extension has been demonstrated to provide a substantial benefit to the glass manufacturer in the management of existing assets. In addition, it has been proven in other applications to be very adaptable to installations in oxy-fuel furnaces and air-fuel furnaces. Its operation can be for zero port boost applications or extended to supply over 50% of the total energy going into an air-fuel furnace.

These applications illustrate only what has occurred in the short time since BOC commercialized the technology. The future of CGM appears to be as exciting as its short history. The U.S. Department of Energy commissioned the Glass Manufacturing Industry Council, who contracted with industry consultants P. Ross and G. Tincher, to complete a technical and economic assessment study of past glass melting technologies. This report provided the industry with some guidance on the future of the glass melting. One of their conclusions was that the glass industry supported the concept of segmenting or modularizing the glassmaking process in the future. This process focuses on the realization that melting is a different process than refining and may involve different technologies to achieve improved performance. Figure 10 illustrates the adaptability of CGM to a segmented or modular melter.

CGM applied to a separate, high-performance melter, in association with other mixing technologies, such as oxygen bubbling or small submerged combustion burners, has the capability of providing stone-free glass from a very small, cost-effective system.

Summary

BOC's Convective Glass Melting (CGM) technology has proven effective in float furnaces in a zero port boost or hybrid configuration as a furnace life extender or load-recovery agent. At Pilkington's commercial facility, using the CGM system resulted in 51 000 tonnes of load recovered in a span of 14 months.

The oxygen cost associated with a boost can be managed by making an

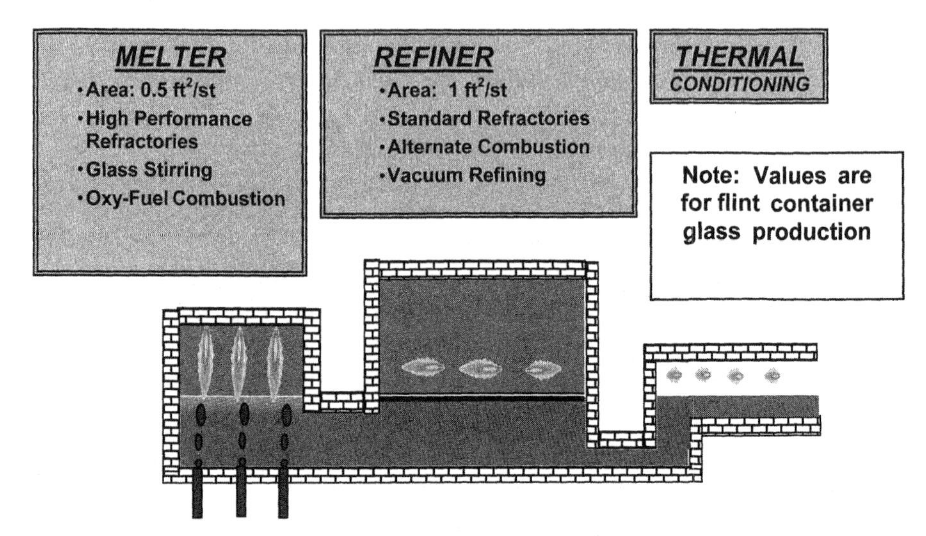

Figure 10. Adapting CGM for a future glass-melting concept.

informed decision on the mode of supply. The analysis indicates that installing an on-site supply toward the end of the furnace campaign provides the glass manufacturer full use of the asset for 2–3 years at the onset and enough flexibility to match production with demand in the new campaign. This supply mode will be superior to liquid supply in certain cases.

While CGM has proven to deliver value in today's furnace designs, it is also a glass-melting technology that can be adapted to the future generation of segmented or modular glass melters.

Fire Polishing with Premixing Technology

Hans Mahrenholtz
Linde AG, Linde Gas Division, Unterschleissheim, Germany

Different heat transfer rates, flame shapes, working parameters, and so on for premixed and post-mixed oxy-hydrogen and oxy-methane flames result in differences between the flames. The heat transfer rates from premixed oxy-hydrogen flames are approximately six times higher than those from premixed oxy-methane flames. The heat transfer rates from premixed flames are approximately 30% higher in the case of oxy-methane flames and approximately two times higher in the case of oxy-hydrogen flames than in post-mixed flames. The heat transfer profile in radial distance of oxy-hydrogen flames is sharper than in oxy-methane flames. The flame width for oxy-hydrogen flames is smaller than for oxy-methane flames, assuming that the flame width in this case is defined by heat transfer rates lower than 150 kW/m^2. This means that premixing glass treatment technologies are more efficient than post-mixing technologies and that glass treatment technologies with hydrogen/oxygen are more efficient than with natural gas/oxygen or propane/oxygen. When it comes to meeting the highest demands in glass manufacturing — especially high glass quality and short working times — only premix technology with oxygen and hydrogen brings the requested results.

Introduction

In the glass industry — in tableware and flacon production in particular — the demands being placed on quality (especially on glass surface quality) and on production costs are increasing to an equal degree.

Today, modern glass production demands:

- The highest glass quality, particularly the highest surface quality.
- Machined glass to look like manual glass.
- The replacement of acid polishing.
- The elimination of further mechanical processing.

The difference between premixing and post-mixing burners is shown in Fig. 1. Post-mixing burners are quite simple. Each burner has two inlets, one for gas and one for oxygen/air. These two media are led separately within the burner and mixed at the surface of the burner through a system of gas and oxygen/air nozzles. After the mixing, the flame/combustion starts.

Premixing burners work differently. Gas and oxygen/air is mixed at a mixing station and then the gas mixture is delivered to the burner. The burner has only one inlet, and after passing the nozzle system the combustion starts.

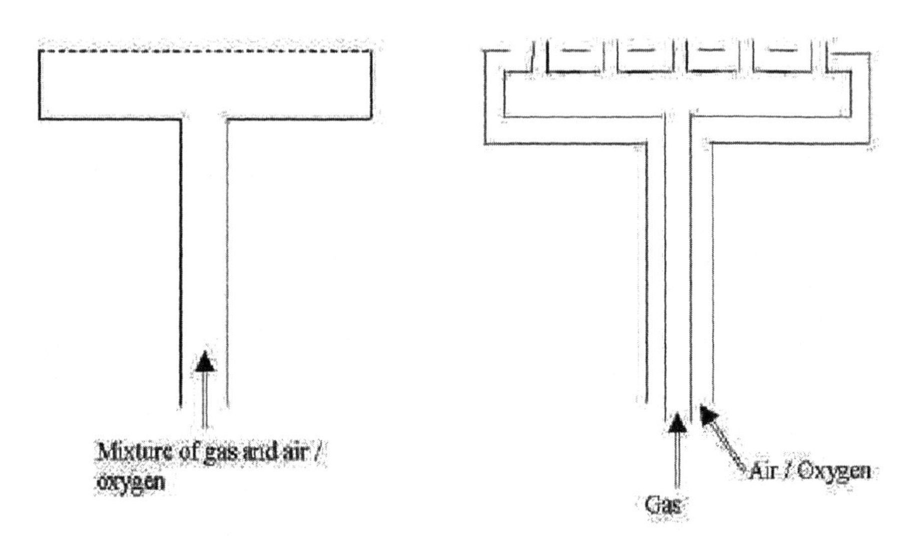

Figure 1. The difference between premixing and post-mixing burners.

Heat Transfer Rates for Premixing and Post-Mixing Burners

Axial Distribution of the Heat Transfer

The axial heat transfer rates for premixing and post-mixing burners are shown in Fig. 2. The combustion of premixed oxy-hydrogen starts immediately after the nozzle system, whereas post-mixed oxy-hydrogen begins to burn a little further than 2 cm. The maximum heat transfer rate for premixed oxy-hydrogen is about 1550 kW/m^2 slightly below 2 cm far from the burner nozzles; for post-mixed oxy-hydrogen it is about 800 kW/m^2 at nearly 3 cm from the burner nozzles.

The combustion of premixed oxy-methane starts around 4 cm, whereas the post-mixing variety starts to burn further than 7 cm. The maximum heat transfer rate for premixed oxy-methane is about 270 kW/m^2 close to 5 cm from the burner nozzles; for post-mixed oxy-methane it is about 180 kW/m^2 and 7 cm from the burner nozzles.

This means that in particular that the heat transfer rates for premixed

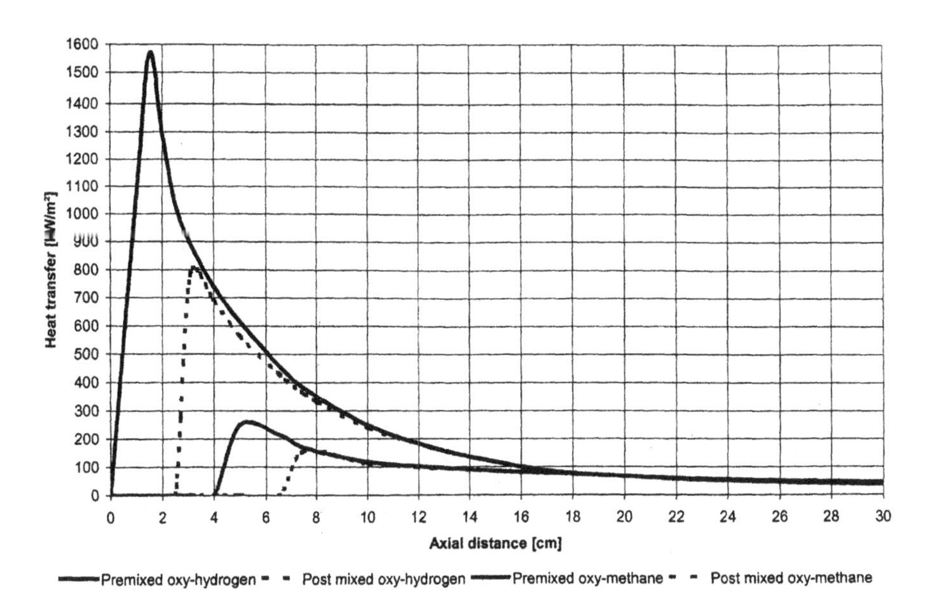

Figure 2. Axial distribution of the heat transfer for oxy-hydrogen and oxy-methane flames.

oxy-hydrogen flames are approximately six times higher for than premixed oxy-methane flames, and that the heat transfer rates from premixed flames are approximately 30% higher in the case of oxy-methane flames and approximately two times higher in the case of oxy-hydrogen flames than for post-mixed flames.

Radial Distribution of the Heat Transfer

The radial heat transfer rates for premixing burners are shown in Fig. 3. At the point of the maximum heat transfer in the axial direction, the premixed oxy-hydrogen flame has a heat transfer rate of 50 kW/m^2 at a radial distance of slightly lower than 3.0 cm and 100 kW/m^2 at a radial distance of about 2.75 cm.

The premixed oxy-methane flame has a heat transfer rate of 50 kW/m^2 at at a radial distance of nearly 6.0 cm and 100 kW/m^2 at a radial distance of 4.5 cm.

This means in particular that the heat transfer profile in radial distance of oxy-hydrogen flames is sharper than for oxy-methane flames, and that the

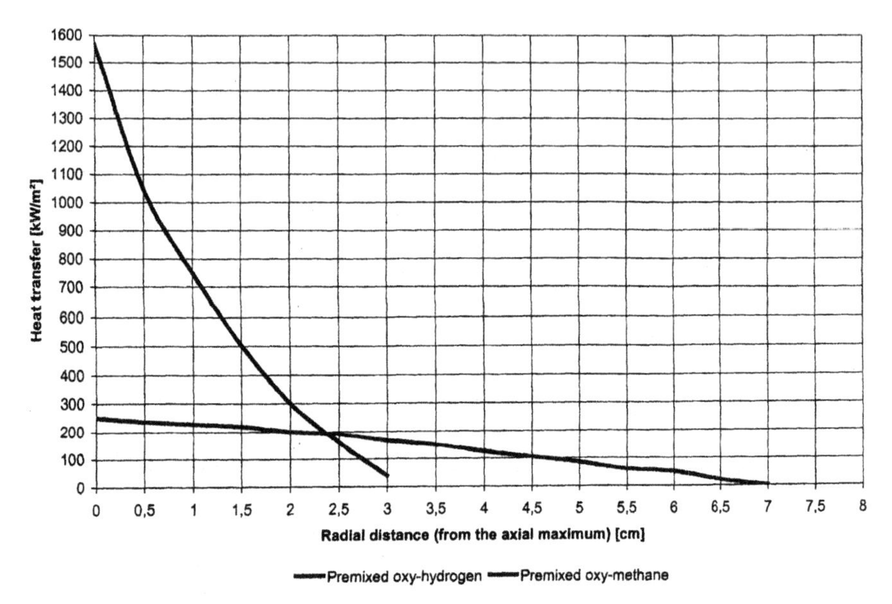

Figure 3. Radial distribution of the heat transfer for oxy-hydrogen and oxy-methane flames.

flame width for oxy-hydrogen flames is smaller than for oxy-methane flames, assuming that the flame width in this case is defined by heat transfer rates lower than 150 kW/m^2.

Flame Basics

To explain the above results some flame and combustion basics have to be discussed. This discussion will focus mainly on flame front equilibrium, entrainment, and heat transfer mechanisms.

Flame Front Equilibrium

The flame front is the point where the medium velocity and the flame velocity are equal (and heat transfer starts). The medium flow wants to push the flame front downstream while the flame wants to burn upstream.

The flame velocity is a function of several properties (density, viscosity, etc.) and is proportional to the diffusion coefficient (Eq. 1) and the reaction velocity/coefficient (Eq. 2).

$$v_F \sim k \tag{1}$$

$$k = k_0 \cdot e^{-\frac{\triangleleft E}{R \cdot T}} \tag{2}$$

Because the activating energy for oxy-hydrogen combustion is lower than for oxy-methane (Eq. 3) the diffusion coefficient for oxy-hydrogen is higher than for oxy-methane (Eq. 4) and in the end the flame velocity for oxy-hydrogen flames is also higher than for oxy-methane (Eq. 5).

$$\triangleleft E_{oxy-hydrogen} < \triangleleft E_{oxy-methane} \tag{3}$$

$$k_{oxy-hydrogen} > k_{oxy-methane} \tag{4}$$

$$v_{oxy-hydrogen} > v_{oxy-methane} \tag{5}$$

This means that the medium flow velocity for oxy-hydrogen can be much higher than for oxy-methane, with having a flame. The maximum speed for premixed oxy-hydrogen flames is about 3–4 times higher than for premixed oxy-methane flames.

Entrainment

As a consequence of momentum exchange between the jet and its surroundings, fluid enters from the surroundings across the boundaries of the jet. This entrainment is due to friction, which depends upon the exchange coefficient and the velocity gradient. The presence of turbulence increases the exchange coefficient several orders of magnitude in respect to the molecular exchange coefficient. In premixed jet flames the entrainment of air into the jet has two effects:

1. Before ignition, the oxygen entrained from the surrounding air changes the oxygen/fuel ratio of the combustible mixture, bringing the mixture to over-stoichiometric conditions. The entrained nitrogen dilutes the mixture, decreasing the concentration of fuel and oxygen.
2. After ignition, the air-entrained oxygen may participate in combustion of unburned fuel; however, the main effect of the entrainment is the reduction of jet temperature.

Heat Transfer/Flux

Very often the effective heat is transferred by radiation and convection, and because there is interference/interaction the effective heat transfer is the sum of those two (Eq. 6).

$$\dot{q} = \dot{q}_{\alpha} + \dot{q}_{\varepsilon} \tag{6}$$

Heat Transfer by Radiation

Heat transfer by radiation can be described by the following well-known estimation according to Planck's and Stefan-Boltzmann's law (Eq. 7).

$$\dot{q}_{\varepsilon} = k \cdot \varepsilon \cdot \sigma \cdot T^4 \tag{7}$$

With a certain wavelength the emission coefficient is equal the adsorption coefficient according to Kirchoff's law (Eq. 8). With Arrhenius (Eq. 9), the adsorption is a function of several variables, especially the layer thickness. In the case of glass treatment technologies this layer thickness is the distance between the nozzle, the flame, and the workpiece.

$$\varepsilon_{(\lambda)} = A_{(\lambda)} \tag{8}$$

$$A_{(\lambda)} = 1 - e^{(-a \cdot p \cdot s)} \tag{9}$$

With a distance of less than 10 cm (3 cm for premixing oxy-hydrogen technologies) the adsorption is nearly zero (Eq. 10), which means that the heat transfer by radiation is practically zero (Eq. 11).

$$s \rightarrow 0 \Rightarrow A_{(\lambda)} \rightarrow 0 \tag{10}$$

$$\dot{q}_{\varepsilon} = 0 \text{ and } \dot{q} = \dot{q}_{\alpha} \tag{11}$$

Heat transfer by convection (impingement) is the main mechanism of heat transfer. The contribution from radiation is negligible.

Heat Transfer by Convection

The convective heat transfer is given by Newton's law (Eq. 12):

$$\dot{q}_{\alpha} = k \cdot \alpha \cdot \triangleleft T \tag{12}$$

The heat transfer coefficient is a function of Nusselt, which is a function of Reynolds and Prandl numbers (Eqs. 13 and 14).

$$\alpha = f(Nu), \quad Nu = f(Re, Pr) \tag{13}$$

$$Re = \frac{\nu_o \cdot d_h}{\nu} \tag{14}$$

Because the velocity for oxy-hydrogen is higher than for oxy-methane (Eq. 15), the Reynolds number for oxy-hydrogen combustion is higher than for oxy-methane (Eq. 16), and in the end the heat transfer coefficient for oxy-hydrogen combustion is also higher than for oxy-methane (Eq. 17).

$$\nu_{oxy-hydrogen} > \nu_{oxy-methane} \tag{15}$$

$$\Rightarrow Re_{oxy-hydrogen} > Re_{oxy-methane} \tag{16}$$

$$\Rightarrow \alpha_{oxy-hydrogen} > \alpha_{oxy-methane} \tag{17}$$

Characteristics of Lindes Premixing Technologies

The main characteristics of Lindes premixing technologies are:

- Premixing of the gas mixture.
- Electropneumatic control units.
- Water-cooled or uncooled burners.
- Standard burners or specially designed burners.
- 100% stainless steel.
- Power control, control units, exact pressure regulation, and exact regulation of the gases ratio for each burner.

Table I is a short overview of premixing oxy-hydrogen and oxy-methane. The advantages and typical uses of premixing oxy-hydrogen include:

- Complete removal of pressing burrs and sharp edges such as stern seams, cover ring seams, and plate seams on thin-walled glasses (particularly glasses with stems); number of cuts up to 60/min.
- High-grade finishing of flacons (brilliance).
- Preventing deformation of the article by using a very short polishing time of approximately 2–3 s.

Table I. Characteristics overview of Hydropox and Carbopox

	Hydropox	Carbopox
Gases used	Hydrogen plus oxygen	Natural gas/propane plus oxygen
Typical areas of application	Fire-polishing of thin machine-pressed or blown glass articles	Fire-polishing of thick machine-pressed or blown glass articles
Working distance	~10–20 mm	~40–100 mm
Run time	~2–5 s	~5–10 s

- Removal of cold waves.
- Heating at precise points.
- Polishing decorative surfaces with very deep reliefs is possible.

The advantages and typical uses of premixing oxy-methane include:

- Removal of pressing burrs and sharp edges.
- Removal of orange peel on pressed lead crystal articles.
- Significantly improved brilliance compared with acid polishing.
- Great cost savings compared with acid polishing.
- Complete polishing of thick-walled glass parts (external and internal).
- Preventing deformation of the article through fast polishing (polishing time approximately 5–15 s).
- Removal of cold waves.
- Polishing decorative surfaces with very deep reliefs is possible.

Figure 4 shows typical designs for Linde premixing burners and line and ring burners. Figure 5 shows the independence of an L-line burner from the position of the gas mixture inlet. The advantage is that it is possible to polish relatively high-glass articles (e.g., vases) effectively.

Equipment and Sample Installations

Each burner is equipped with power control, control units, exact pressure regulation, and exact regulation of the gas ratio.

A complete gas mixing station can look like that shown in Fig. 6.

A burner steering system contains an electrical operating unit, a central mixing unit, and electropneumatic burner controlling system, and a burner.

A principle flowsheet is shown in Fig. 7.

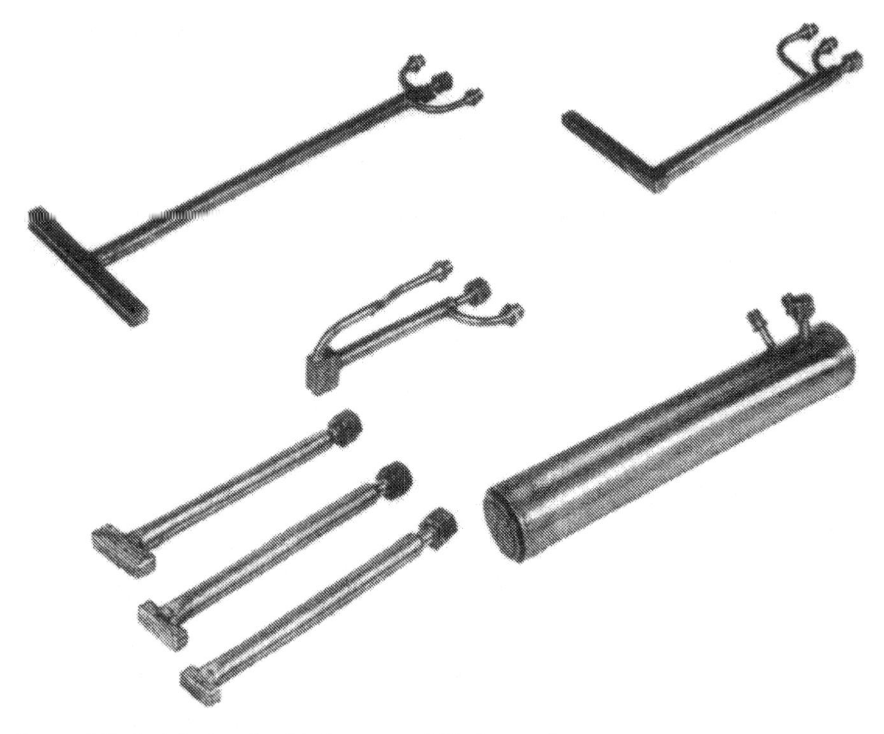

Figure 4. Typical designs for Hydropox and Carbopox burners.

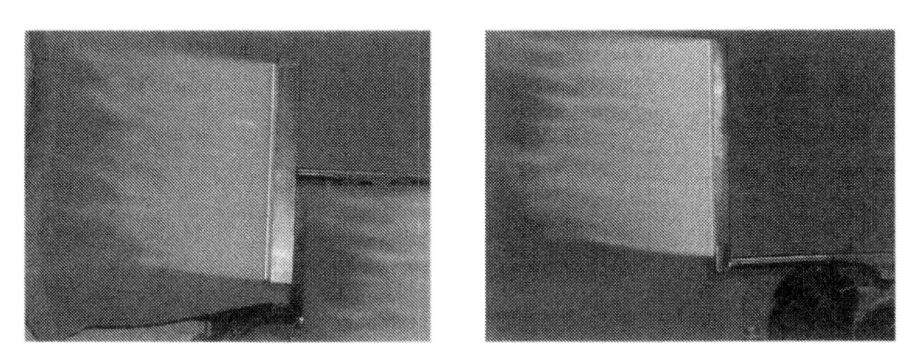

Figure 5. Flame shape of (left) T and (right) L versions of a Carbopox burner.

Figure 6. Gas mixing unit for 2×12 burners.

The complete installation of Carbopox/Hydropox guarantees exact and reproducible conditions for each glass product, and matured installation with a maximum of safety.

Figures 8, 9, and 10 show typical installations.

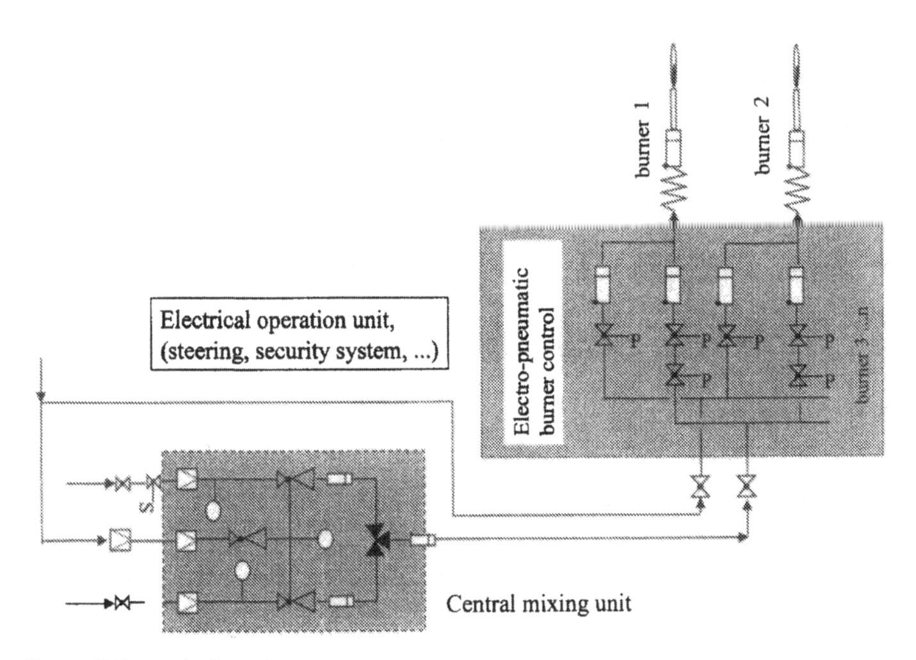

Figure 7. Principle flow sheet.

Figure 8. Surface polishing with Carbopox.

Figure 9. Fire polishing after the forming process with Hydropox.

Figure 10. Total removal with Hydropox.

A Novel Glass Furnace Combining the Best of Oxy-Fuel and Air-Fuel Melting

Mark D'Agostini, Michael E. Habel, Russell J. Hewertson, Bryan C. Hoke Jr., Richard Huang, Julian L. Inskip, Kevin A. Lievre, and Aleksandar G. Slavejkov
Air Products and Chemicals, Inc., Allentown, Pennsylvania

A novel furnace concept has been developed that combines oxy-fuel combustion over the unmelted batch and air-fuel combustion downstream. This hybrid furnace offers a number of advantages over both air-fuel and oxy-fuel furnaces. As compared to air-fuel, the hybrid furnace technology provides increased production, improved glass quality, fuel savings, better furnace temperature control, and a more stable batch pattern. As compared to 100% oxy-fuel, the hybrid furnace delivers similar production levels, improved glass quality, reduced levels of foam on the glass surface, and lower overall oxygen costs. One can think of the hybrid furnace as the optimum between air-fuel and oxy-fuel melting. The end result is that the hybrid furnace provides the means to achieve the lowest overall glass melting cost.

From the emissions point of view, combining oxy-fuel and air-fuel combustion in a furnace presents a challenge with regard to NO_x emissions. An in-depth modeling study was undertaken to overcome this issue with excellent results. Burner design considerations for implementing the hybrid furnace will be presented. A discussion of a detailed economic model comparing the costs of hybrid furnace technology relative to air-fuel and full oxy-fuel technology is included. The results are discussed in terms of the sensitivity to parameters such as production increase, yield, and fuel usage.

Introduction

The benefits of using oxygen for combustion in glass making are well understood. By replacing air (which contains almost 80% inert nitrogen) with pure oxygen, the efficiency of combustion is increased substantially. The temperature of an oxy-fuel flame is significantly higher than that of an air-fuel flame. The dominant heat transfer mode in combustion-based glass melting is radiation. The amount of heat transferred via radiation increases to the fourth power of increases in flame temperature.

The effect of reducing the amount of nitrogen going into the furnace can be clearly seen in Fig. 1, which shows the energy carried out of the furnace for three distinct cases: air-fuel with no heat recovery, air-fuel using regenerators, and full oxy-fuel.

An air-fuel system with no heat recovery loses 81% of the heat input to the furnace in the form of hot gases exiting the furnace. This leaves only 19% of the heat available for melting the glass and overcoming other heat

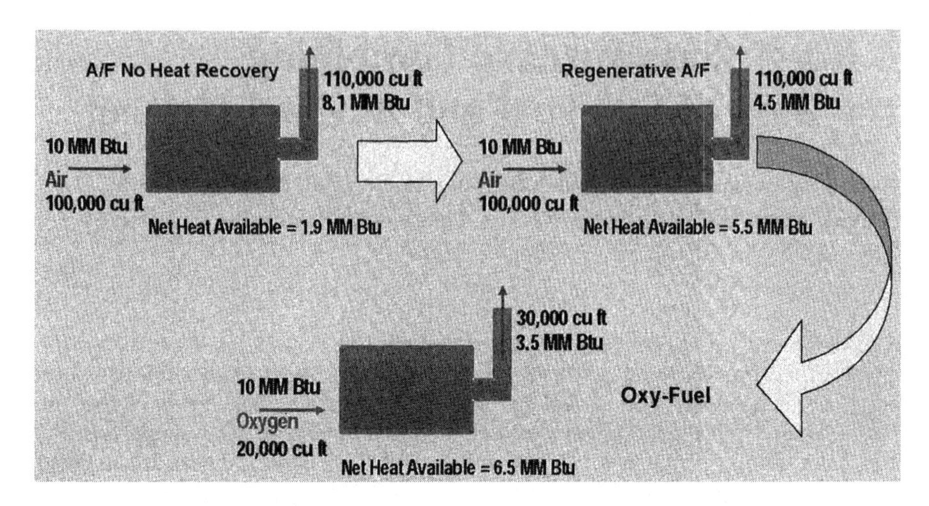

Figure 1. Available heat for three different operating parameters.

losses associated with the furnace (i.e., through refractories, water-cooled jackets, air-cooled equipment, and furnace openings). The addition of a heat recovery device, such as a regenerator or recuperator, reduces the amount of heat loss by preheating the air used in the combustion process. While this is a great improvement, 45% of the heat input to the furnace is still lost to the atmosphere in furnaces of this design. For both of the air-fuel cases described, there are very high temperatures and high nitrogen concentrations in the furnace, leading to high NO_x emissions.

In oxy-fuel furnaces the temperatures are also high, but the scarcity of nitrogen means that very little NO_x is formed. With oxy-fuel, the volume of the flue gases is approximately 20% of the air-fuel cases, so that the amount of heat carried out with these gases represents only 35% of the heat input. As compared with a regenerative furnace, the amount of heat available for melting the glass is increased by almost 20%. Figure 2 shows the effect of oxygen enrichment on total available heat for a fixed flue gas temperature.

The earliest use of oxygen in the melting chamber was through global enrichment of the combustion air. In this application, the oxygen content can be raised from 21% up to levels as high as 30%. This technique is used widely in cases where regenerators are clogged or have collapsed and air-flow is reduced. A more selective technique for the introduction of oxygen is referred to as "lancing." Lancing is the practice of strategically injecting oxygen through a pipe into the combustion zone. This technique is used when oxygen is needed in localized sections of the furnace, for example, to

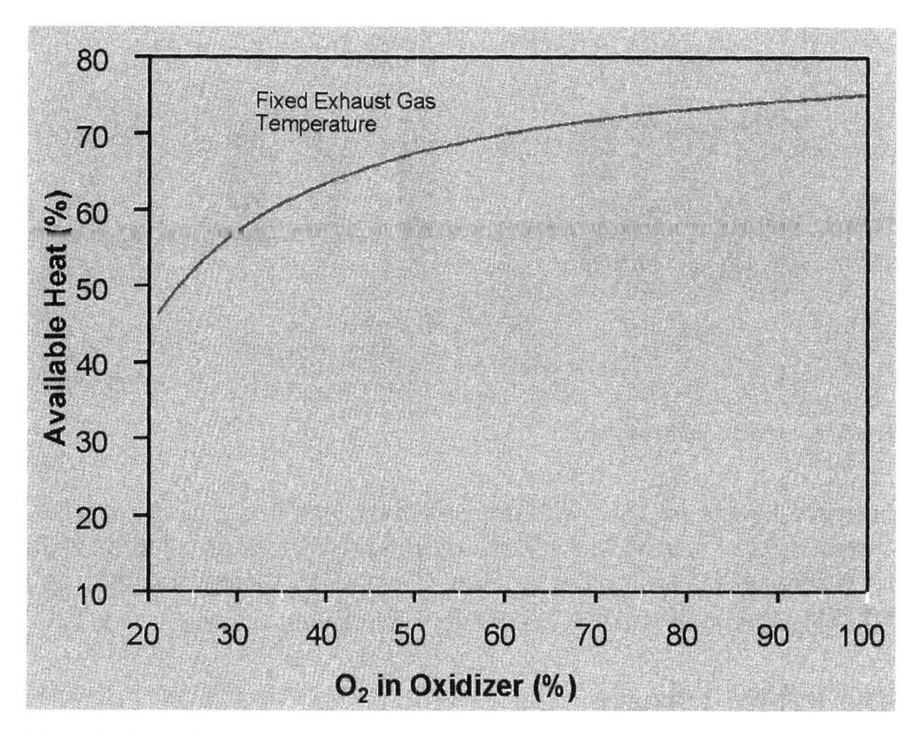

Figure 2. Available heat versus O_2 in oxidizer.

selectively enrich only those ports where air flow is reduced because of partially plugged checkers.

A comparatively more advanced method for the introduction of oxygen involves the use of oxy-fuel combustion. The practice of adding oxy-fuel burners to a system designed for air-fuel operation is referred to as supplemental oxy-fuel, or oxy-fuel boosting. These burners are often located close to the batch charging area to maximize melting efficiency.

A full oxy-fuel furnace represents the most thermally efficient fossil fuel–based glass melter commonly used in the industry. This technology uses specially designed oxy-fuel burners and flow control systems, incorporates different furnace design with respect to burner and flue locations, and eliminates heat recovery devices such as regenerators and recuperators. Full oxy-fuel firing in glass furnaces is a proven technology. Currently, over 7 million tons of glass per year are melted using oxy-fuel.

Air Products's understanding of the strengths and weaknesses of air-fuel and oxy-fuel led us to the introduction of our latest technology, which com-

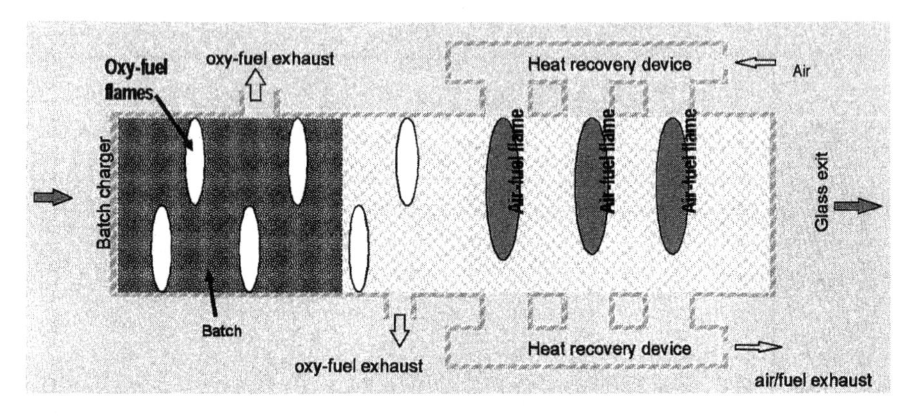

Figure 3. Hybrid furnace schematic.

bines the best of oxy-fuel with the best of air-fuel. It is a measure of our customer focus that while Air Products is an industrial gas company with an interest in selling oxygen, we have developed technology that uses significantly less oxygen than a full oxy-fuel conversion.

The Hybrid Furnace

The hybrid furnace technology uses oxy-fuel combustion in the batch charge end of the furnace and air-fuel with heat recovery in the down-tank refining section of the furnace. Figure 3 shows this furnace design as described by Hoke et al. (2002). To understand why this makes sense it is helpful to study the relative strengths of oxy-fuel and air-fuel.

Air-Fuel Relative Strengths

The biggest advantage of air-fuel combustion is that air is available at low cost. As shown in Fig. 1, the use of heat recovery allows air-fuel systems to be fairly efficient, considering that 80% of the air is inert. Another key advantage to air-fuel systems is the operational experience that has been built up over many furnace campaigns. Recurring problems have been solved and refractory performance and characteristics are well understood. As furnaces get larger and campaign targets get longer, this know-how plays an important role in minimizing risk.

Some glassmakers have reported increased foam in the refining area when using oxygen as compared to air-fuel.[1] It is not clear whether more foam is formed with oxygen firing, or if the higher velocities of the air-fuel combustion gases combined with constant switching of firing from one side to the other simply break up the foam more effectively. Foam acts as an insulator, which reduces heat transfer to the glass, potentially increasing refractory temperatures and crown wear. A factor in foam formation is localized overheating or reboil. This occurs when heat release is localized and glass is overheated. In the hybrid operating system, the downtank energy is provided via air-fuel combustion. The larger volumes of oxidizer and fuel from air-fuel combustion (required to deliver the same heat as compared with oxy-fuel) can be used to more effectively spread the heat release in this critical area of the furnace and thereby minimize foam formation. Air Products has conducted furnace trials that have shown the effectiveness of air-fuel and air-oxy-fuel systems in breaking up foam in oxy-fuel furnaces.

Oxy-Fuel Relative Strengths

As discussed earlier, oxygen combustion is more efficient than air-fuel because of the elimination of nitrogen. Air-fuel efficiencies can be increased by using recuperators and regenerators, but this adds costs, maintenance requirements, and complexity to the glassmaker's operation and efficiencies still fall short of oxy-fuel furnace operations. The higher efficiency of oxy-fuel leads to fuel savings, which can be a significant driver considering today's high fuel costs.

Another advantage of oxy-fuel is the simplicity of the combustion system. Heat recovery is typically not used because of the much lower heat losses to be recovered. This eliminates problems with regenerators and recuperators such as blockage of ports (particularly near the batch charger), checker collapse, uneven flow rates, and reduced campaign life. In addition, oxy-fuel systems fire at a continuous rate, controlled by the glassmaker without interruption. The benefit can be clearly seen from the stable batch patterns achieved in oxy-fuel furnaces when compared to air-fuel furnaces.

Oxy-fuel burner technology has delivered impressive gains in productivity and glass yields. Numerous articles have been published about production increase and yield improvement when converting a furnace from air-fuel to oxy-fuel combustion.[2,3] State-of-the-art flat flame burners, such as the Cleanfire HR burner, have shown significant improvements in melt rates and glass quality. By maximizing the radiation achieved, spreading it over a large flame area, and directing it down to the melt, the batch is melted more quickly, using less fuel. The size of the refining zone and residence time for glass in that zone are both increased with the net result being extra production and higher quality. In any economic evaluation, these two factors are vitally important and provide a significant advantage to the glassmaker.

The environmental benefits of oxy-fuel are often the driver for a glassmaker converting to oxy-fuel. By virtually eliminating nitrogen from the furnace, NO_x levels can be exceedingly low with proper furnace design and burner selection. Reductions of 10–20 times compared to regenerator furnaces have been demonstrated. Another environmental benefit is the reduction of batch carryover. The quicker melting and the lower gas velocities, combined with different flue arrangements, mean that far less batch is blown out of the furnace, thus reducing losses, handling costs, and emissions. Batch carryover into the early ports of regenerative furnaces is one of the biggest ongoing challenges for glassmakers.

Float furnaces offer a special case for oxygen supply. The tin bath used for forming the glass ribbon needs large quantities of nitrogen to prevent the tin at the surface from oxidizing. When the decision is made to convert the furnace to oxy-fuel, an air separation plant can be designed to produce both oxygen and nitrogen more efficiently than supplying each gas individually. This is achieved by taking advantage of equipment and process synergies in the plant design. So, float customers who need nitrogen and oxygen can achieve a lower overall cost for the two gases.

Hybrid: The Best of Air-Fuel and Oxy-Fuel

Combining the proven techniques of air-fuel melting and oxy-fuel melting in a way that delivers the upsides of each, while overcoming the inherent challenges of using both systems in the same furnace, requires close attention to detail and in-depth knowledge of glassmaking. The next section describes the benefits, challenges, and results of the hybrid system.

Benefits

Glass Quality

Using oxygen in the melting zone maximizes the melt rate and fuel efficiency. The perennial problem of plugging or collapsing in the up-tank ports because of batch material carryover in air-fuel operations is eliminated when oxy-fuel firing is used in this region of the furnace. The continuous firing in the melting area delivers stable batch patterns, and because the batch melts early, the refining section is effectively made larger. This

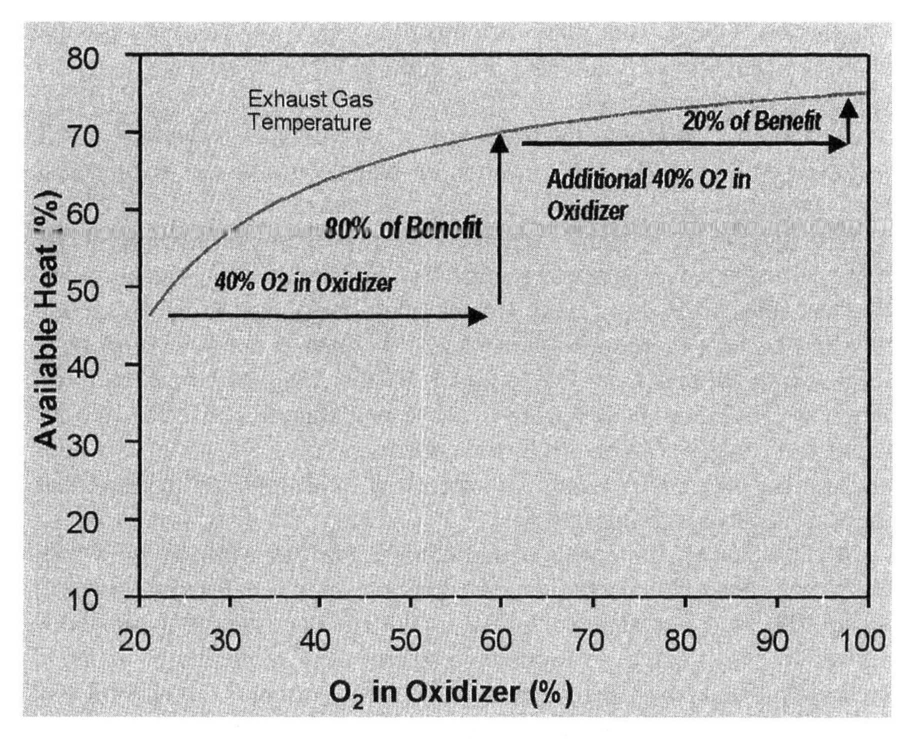

Figure 4. Available heat: degree of benefit versus O_2 in oxidizer.

expansion of the refining zone increases glass particle residence times, leading to higher quality glass. Whereas the stability of the oxy-fuel in the melting area helps deliver good quality, paradoxically it may be the inherent instability of the regenerator system in the refining zone — with the variation of flame temperatures, periods with no firing, and the reversal of the flames — that minimizes foam and further improves glass quality.

Furnace Efficiency

As outlined earlier, available heat is the energy available to the process when the energy lost in the flue gases is accounted for. The available heat curve shown in Fig. 4 highlights one of the key drivers for the hybrid furnace concept: It is the first amounts of oxygen added that deliver the biggest increase in available heat and therefore efficiency. For this temperature, transitioning from air at 21% oxygen to 60% oxygen enrichment level increases the available heat from 46 to 70%, providing 52% more heat to the process. Going from 60% enrichment to 100% oxygen increases the available heat from 70 to 75%, only adding 7% more heat for a similar change in enrichment level.

Air Products designs combustion systems for multiple industries. Often, solutions developed for one industry can be customized and fine-tuned to solve problems in another. For example, the variation of available heat is widely understood in many industries, such as aluminum melting, where highly efficient rotary furnaces use full oxy-fuel because they can absorb all the heat effectively, but reverb (rectangular, stationary) furnaces use air-oxy-fuel burners because heat transfer to the bath is not quick enough to fully absorb the heat from full oxy-fuel. By matching the heat input to the inherent efficiency of the process, these manufacturers use the optimal amount of oxygen. The reverb furnace operators achieve lower overall costs and use less oxygen by using a combination of air-fuel and oxy-fuel than they can achieve with full oxy-fuel.

In glassmaking, low levels of enrichment, such as with oxy-boost systems, are an established example of delivering significant benefits with only about 10% of the total furnace heat input. The operation is optimized to deliver the maximum glass production benefit while minimizing the investment capital and power requirements of the oxygen supply. The hybrid concept is really a matter of using just the right amount of oxygen for your needs — no more, no less.

Hot and Cold Furnace Repairs

All too frequently, failure of the regenerators (particularly ports 1 and 2) limits the furnace campaign or reduces a furnace's capacity and glass quality. Because most arrangements for hybrid technology eliminate the requirement for healthy checker packs in this region of the furnace, it is an excellent fit to repair furnaces mid-campaign and to bring them back to their design capacity, or even better. For this purpose, we have developed a technology that will enable the glassmaker to install state-of-the-art oxy-fuel burners within the ports that deliver a flame with optimal properties: luminous, with excellent geometry, stability, and melter area coverage providing efficient heat transfer directly above the glass.

For furnace rebuilds or greenfield sites, the hybrid furnace reduces refractory material expenses, installation labor, and downtime. Careful furnace design, burner selection, firing patterns, and flow control will deliver many of the benefits of full oxy-fuel with less oxygen. The advantage highlighted earlier for float furnaces of producing both oxygen and nitrogen is further refined for hybrid technology. This is because an oxygen plant built to provide oxygen for a full float furnace conversion can be designed, with low incremental cost, to make enough nitrogen for two tin baths. A further benefit for multiple furnace sites is that because much less oxygen is required when operating with the hybrid furnace, the same air separation plant that would previously have been used to supply the gases for a single full oxy-fuel conversion is now able to provide enough nitrogen and oxygen for two float furnaces using the hybrid concept.

Oxygen Supply Considerations

A frequent comment from glassmakers is that the cost of oxygen is too high. By significantly reducing the amount of oxygen used, the total cost can be greatly reduced. Along with this reduction in oxygen requirements comes the benefit of improved reliability and accessibility to oxygen technologies for glassmakers in areas where industrial gases have not previously been viable. The typical supply scenario for a manufacturer with a large-scale glass furnace is to have an air separation unit generate oxygen at the site. While these generators operate reliably 99% of the time, provisions must be made to supply oxygen when the generator is down. This backup system usually consists of tanks of liquid oxygen sized to provide enough oxygen for 24–72 h under normal use conditions. If the outage continues for a longer period, then additional oxygen must be trucked in to refill the tanks.

In countries or regions where oxygen supply is not as widespread as, say, the midwestern region of the United States, it may not be economical or may prove very difficult to supply the large quantities of oxygen required for a furnace that is designed and operated completely with oxy-fuel firing during an extended oxygen plant outage. Even in the United States (where the industrial gas distribution infrastructure is mature and established), as more and more furnaces convert to oxygen and supplies tighten, unplanned spot requirements that are made possible by trucking in 200–300 t/day of liquid oxygen to back up the oxygen plant for a full furnace conversion will be much more difficult than supplying the lesser quantities required for a hybrid furnace.

When faced with a complete loss of oxygen, a benefit of the hybrid configuration is that the furnace can be kept hot using the air-fuel system without an emergency effort. In addition to the logistics issues associated with such an emergency effort, full oxy-fuel furnaces sometimes require furnace modifications to allow air-fuel heatup type burners to be used. A further safeguard is provided by the backup applications technologies developed by Air Products. The Cleanfire AOF burner can be quickly installed in a furnace using Cleanfire HR burners to allow air-fuel or air-oxy-fuel firing to be used in the oxy-fuel section of the hybrid melter.

Challenges for Hybrid Technology

As with any new technology there are hurdles to overcome. The primary challenge comes from the fact that the furnace has two distinct sections: air-fuel and oxy-fuel. Furnace geometry must be optimized for the new operating philosophy. Burners in each section must be designed to work in the hybrid environment. Refractories must be designed for the service they will see and the technology installed will vary for hot repair versus cold repair installations.

Operating Philosophy

The combustion space within a hybrid glass furnace comprises elements of air-fuel and oxy-fuel combustion, and the oxy-fuel burner characteristics needed for optimal performance will vary depending upon firing position. Burners situated adjacent to the charge end wall, for example, fire into a relatively quiescent environment and will provide benefit to the extent that they deliver high radiation to the batch without overheating either the end-wall or the crown. Conversely, at the interface between the air-fuel and oxy-fuel sections of the furnace, the combustion space will be much more turbulent. Hence, the oxy-fuel burner flames will need to produce a desirable flame under these conditions.

Air Products's glass and combustion specialists have conducted extensive CFD modeling and laboratory tests that simulate hybrid furnace operating conditions. These model predictions and tests have led to the development of furnace designs, burner technologies, and operating strategies for highly efficient operation of a hybrid glass furnace. In particular, our burner line includes patented enabling technologies for use in either a rebuilt hybrid furnace or an on-the-fly retrofit conversion of an air-fuel melter. Both scenarios of hybrid operation are of practical relevance to glass manufacturers considering adaptation of the technology. On-the-fly conversion during the final months/years of a furnace campaign could significantly extend furnace life by limiting or overcoming regenerator wear, while also

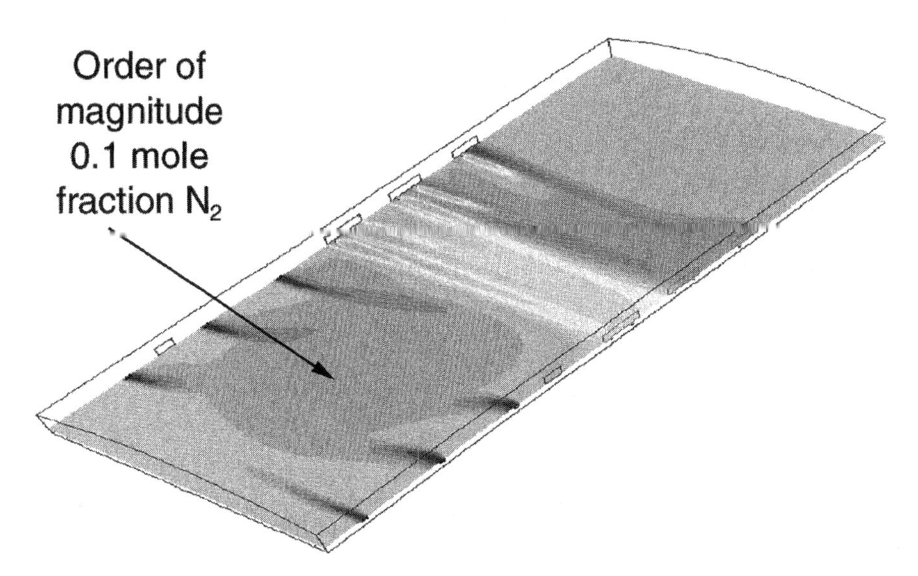

Figure 5. Base N_2 concentrations.

serving as a proving ground for the benefits of hybrid operation. Ultimately, however, a furnace that has been rebuilt to a hybrid design will deliver optimal performance of the technology.

Emissions

Another challenge of the oxy-fuel and air-fuel sections is delivering low-NO_x performance. The high temperature of oxy-fuel flames and the abundance of nitrogen could lead to very high NO_x levels if the furnace layout, firing rates, operating protocol, and so on are not well thought out. Figures 5 and 6 show two cases of the hybrid configuration modeled to predict nitrogen concentrations at the burner plane throughout the furnace. Results achieved by varying structural and operating parameters — such as firing rates, burner locations, flue locations, furnace geometry, and so on — showed a significant difference in N_2 concentrations between the best and worst cases. The key to achieving low NO_x emissions is know-how concerning this technology.

As mentioned earlier, the NO_x emissions from air-fuel systems are often 10–20 times higher than those of oxy-fuel systems. Our studies have shown that a well-designed hybrid case has the potential to deliver NO_x levels 5–10 times lower than air-fuel. While the NO_x levels will probably be higher than for a full oxy-fuel system, a 5–10 times reduction in NO_x will be

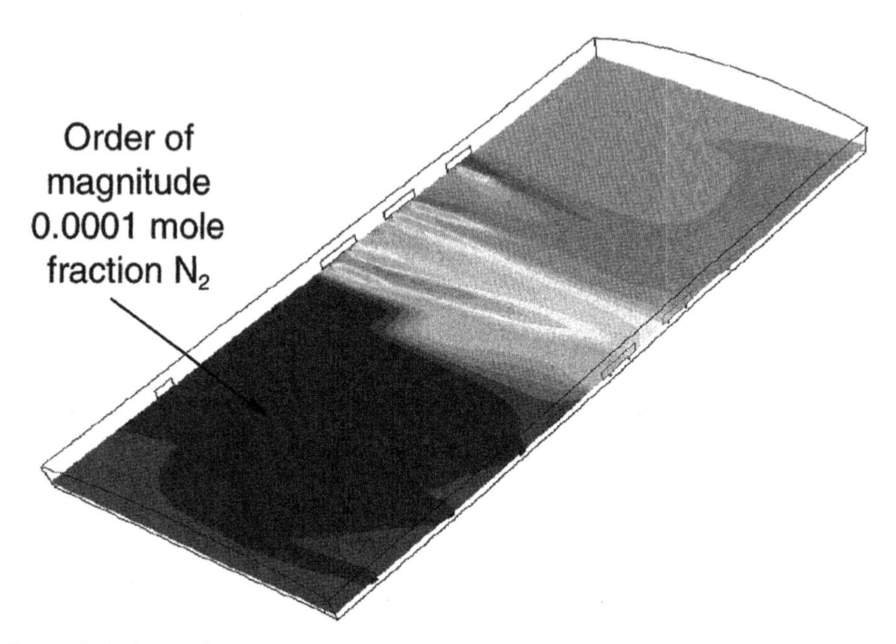

Figure 6. Reduced N_2 concentrations.

very beneficial for many glassmakers. Because fuel usage is decreased compared to air-fuel, the technology has a significant advantage over techniques that use extra fuel to reduce NO_x.

By taking advantage of the decreased fuel consumption and early batch glazing characteristics of the oxy-fuel component within the hybrid system, emissions of carbon dioxide (CO_2) and particulate matter are significantly reduced as compared with traditional air-fuel glass melters. This is especially important as environmental regulations that limit emissions of carbon and carbon-containing species take effect.

Economic Considerations

Air Products has built a detailed economic model for various glassmaking technologies that has shown excellent correlation with data from air-fuel, oxy-fuel, and oxy-fuel boosted air-fuel furnaces. Using this model we project that significant savings can be achieved using hybrid technology. The model includes capital costs such as refractories and labor for construction of the furnace and heat recovery system, as well as the costs associated with the time required for the rebuild, which will be shorter for a hybrid furnace. Operating costs, production levels, fuel usage, and glass yields are also accounted for in detail.

Using the economic model, the hybrid melter provides big benefits when applied to a crippled furnace. Production rates and yields can be greatly reduced and furnace campaigns cut short by premature regenerator failure. The hybrid technology provides heat input where it is most effective, right above the glass, without modifications to the furnace walls or crown. The model projects savings of $5–12 per ton of glass for installations, assuming a production increase of 5–10% and improved yields of 1–2%. Greenfield site installations also show economic benefits when compared to air-fuel, and become all the more attractive when typical end-of-campaign issues are factored into the economics.

Summary

Hybrid furnaces may not be the choice for every situation, but they represent another option for consideration. As compared with air-fuel operations, hybrid technology will deliver fuel savings, production increases, yield improvements, furnace life extension, and emissions reductions. As compared with either air-fuel or oxy-fuel operations, the technology may deliver lower cost, higher quality glass.

Furnaces have been operated with air-fuel combustion systems for thousands of years. This long track record has established familiarity and considerable operating know-how, which weigh in its favor as compared to the relatively new oxy-fuel based technologies. The hybrid melter technology fills the gap between air-fuel and oxy-fuel operating systems. It delivers the best of both technologies, and in many cases will be the lowest cost solution. The next step is to install a full-scale, optimized system and further prove the concept.

References

1. P. R. Laimbock and R. G. C. Beerkens, "Foaming Behaviour and Sulfate Chemistry of Soda-Lime Glass Melts"; pp. 51–53 in *Proceedings Opening TNO Glass Technology Representative Office, North America,* 1998.
2. M. R. Lindig, "TV Oxy-Fuel Conversion and Experience with Noncatalytic Denitrification"; pp. 37–45 in *Proceedings of the 59th Conference on Glass Problems.* American Ceramic Society, Westerville, Ohio, 1999.
3. P. B. Eleazer and A. G. Slavejkov, "Clean Firing of Glass Furnaces through the Use of Oxygen"; pp. 159–174 in *Proceedings of the 54th Conference on Glass Problems.* American Ceramic Society, Westerville, Ohio, 1994.

How Mathematical Modeling Can Help Reduce Energy Usage for Glass Melting

Erik Muijsenberg
Glass Service BV, Maastricht, The Netherlands

Miroslav Trochta
Glass Service Inc., Vsetín, Czech Republic

The state-of-the-art techniques of modeling of glass-melting furnaces give a time averaged solution of the whole (coupled) model. This method seems to be satisfactory to obtain accurate heat fluxes between the glass and combustion space, and thus to have a good boundary condition for the glass model to evaluate all the values of interest, such as glass flow patterns, temperatures, batch shape, quality indices, seed counts, and so on. However, in modeling combustion chambers of regenerative furnaces, we can see processes that, because of their nature, are time-dependent and cannot be represented by a time-averaged steady state. We can, for example, approximate the time-averaged heat fluxes between the combustion chamber and the glass melt even in regenerative furnaces, because we can use symmetry, or, in the worst case, two combustion models that represent firing from each side. However, if we want to simulate the regenerators themselves, we cannot find any time-averaged state. For such purposes, we employ a simplified transient combustion model. Results of simulations of two designs of a regenerator are presented and discussed. Some drawbacks of traditional regenerator designs are identified and improvements suggested.

Time Averaging in Regenerative Furnaces

Modeling studies, which compare two or more cases, are usually based on different steady states of different furnace designs/setups. In reality, of course, there are transient processes in the furnace even during steady conditions. Air preheating by regenerators is a typical example: temperature distribution in the combustion chamber changes with every reversal, as does the heat flux to the glass melt. Thus, we have to find a time-averaged representative of the heat flux.

In many cases, the situation is simple because regenerative furnaces are, by their nature, often symmetrical. All we have to do in such cases is calculate average of the radiative intensity calculated by combustion solver and its mirrored image (see Fig. 1). It more convenient to think in terms of

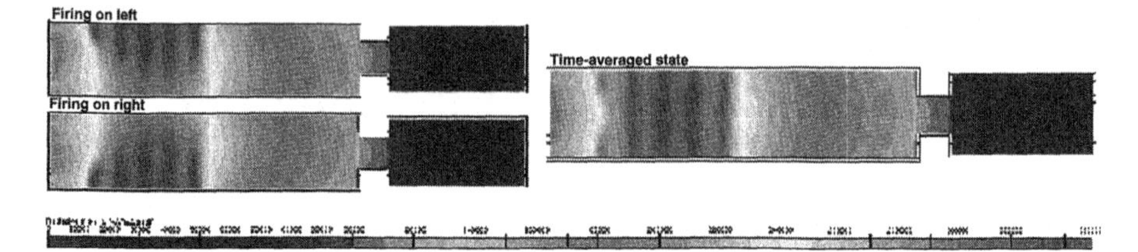

Figure 1. Original radiative intensity on the glass surface, mirrored intensity, and average of both (time-averaged state).

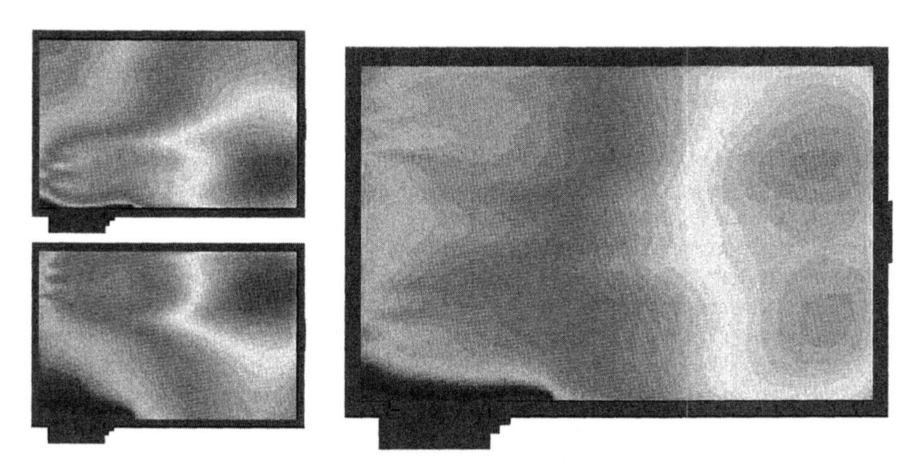

Figure 2. Radiative intensity on glass surface in end-fired furnace (firing from the right, firing from the left, and a time-average created from the two).

radiative intensity instead of heat fluxes because radiative intensity is not directly dependent on the temperature and emissivity of the glass surface.

Some regenerative furnaces are, however, not symmetrical; for example, in end-fired container furnaces, the doghouse and batch chargers are often on one side only. Figure 2 shows a typical example of such furnace. As the batch melts, relatively cold gases (100–1000°C) are released. The gases are mainly CO_2 and H_2O, which are, in addition, good heat absorbers and thus have a strong cooling effect on the crown. This makes the combustion space strongly unsymmetrical and the symmetrization of the heat fluxes is no longer a good approximation. In such cases, we calculate two combustion models (one for firing on the left and one for firing on the right) and the radiative intensity passed to the CM-GM coupling procedure is an average of both.

Modeling Regenerators

Transient Computations

In previous section we described how we obtain time-averaged heat fluxes between the combustion space and the glass melt. This approach allows us to avoid running time-consuming time-dependent computation. Unfortunately, there is no similar time-averaging approach applicable when we want to model regenerators themselves. In that case we should perform a time-dependent computation; however, a simplification is allowed.

The transient terms in transport equations always express temporary instability. For example, the transient term in an energy equation represents release of accumulated heat; the transient term in species transport equations represents the contribution of the accumulated concentration of species. For any of such cases, presence of such a source is temporary and the source will be depleted in certain time horizon.

Residence time of gases in the regenerator is usually on the order of seconds. Most of the phenomena (turbulence, chemistry) are even faster, as is the depletion of sources coming from the transient terms of the equations. What we are interested in is time development in a range of minutes. The only phenomenon that makes the model unsteady in the range of minutes is the heat accumulation in solid regions, especially checkers. This heat is gradually transferred to the fluid via radiative and convective heat transfer so it affects (via temperature-dependent density) all other variables. However, as we mentioned, the other variables reach their steady state much faster so it is not necessary to include the transient terms in other equations except temperature in solid regions (checkers).

The simplification described above is not very dramatic; nevertheless, it saves memory (because we do not have to store values for all the variables in the previous time step[s]) and improves the stability of the computation.

Gettting to Periodical State

As we already mentioned, there is no steady state in the regenerators. However, there is something which we may call "periodical state." That is, temperatures go down during the firing mode (the checkers give energy to the air being preheated) and then go back up to the original level (the checkers receive energy from the hot exhaust gases); this process repeats in every reversal cycle. Unfortunately, it is not easy for the model to get into such a state (in reality, it takes a long time, too).

When we employ transient computation with time step Δt, the solution in time $t + \Delta t$ depends on solution from time t. This means that, unlike in steady computation, we have to start from a state that makes physical sense; in the ideal case, we should start from a state that represents a certain stage of the real process so we can keep track of what is happening in the model. There are many ways; however, the most understandable one is probably to approximate the furnace heatup. Thus, at time t_0, we use a constant temperature of 30°C. Then we start blowing hot combustion gases from the port (we simulate the exhaust side). Also, radiation from the combustion chamber penetrates through the port and heats up the checkers. As the checkers heat up (it takes hours to accumulate the heat), we monitor the temperature of the gases leaving the regenerator. When they reach a value that is close to reality (usually 400–500°C), we consider the regenerator to be heated up and then we start the reversal process: we change the boundary conditions to blow the cold air through the regenerator for a typical reversal cycle half-period, typically 20 min. Then we change the boundary conditions again and blow the hot exhaust gases from top for another 20 min to simulate the exhaust stage of the regenerator. If we repeat this procedure long enough, we should come to the periodical state of the regenerator.

In fact, we do not need to come to the completely periodical state — even the simulation of heating up and then a few reversal cycles will give us a good indication about the performance of the regenerator.

Sample Cases

Inputs

As an example, we chose a regenerator of an end-fired furnace. The overall height of the regenerator is almost 16 m. We decided to compare two designs. The exhaust/air channel in Case 1 is located opposite the port, whereas the exhaust/air channel in Case 2 is on the same side as the port; that is, it is below the tank (see Figs. 3 and 4).

In the firing mode, the air flow is 3.29 kg/s and the input temperature is 30°C. In exhaust mode, there is 3.5 kg/s of combustion gases with a temperature of 1560°C.

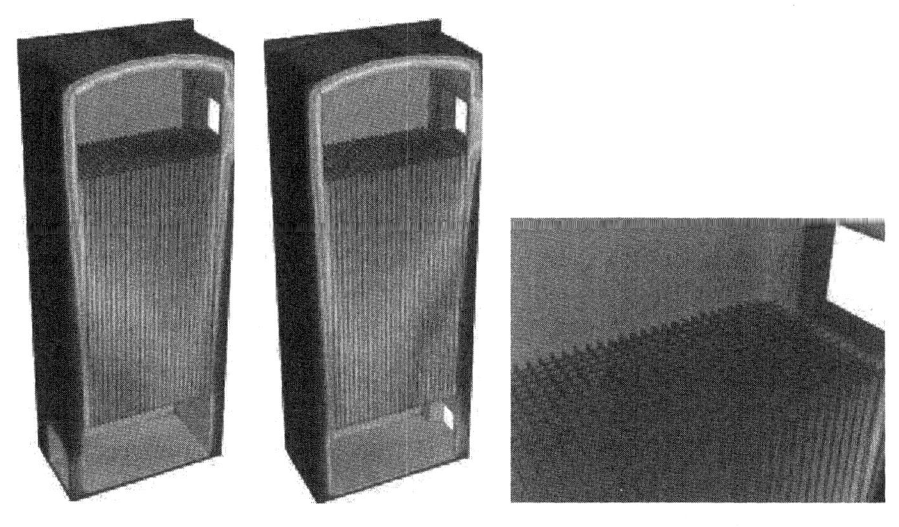

Figure 3. Geometry of Case 1 and Case 2 (shaded by temperature) and a detailed view of the top of the checkerwork.

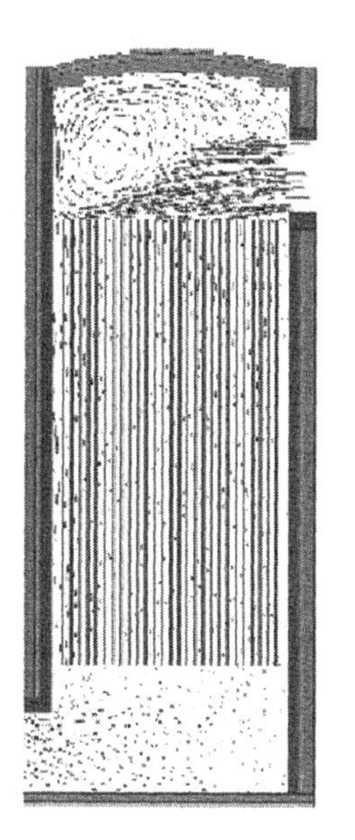

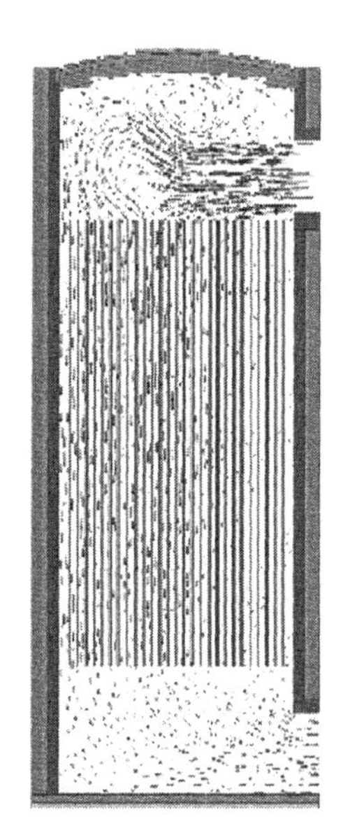

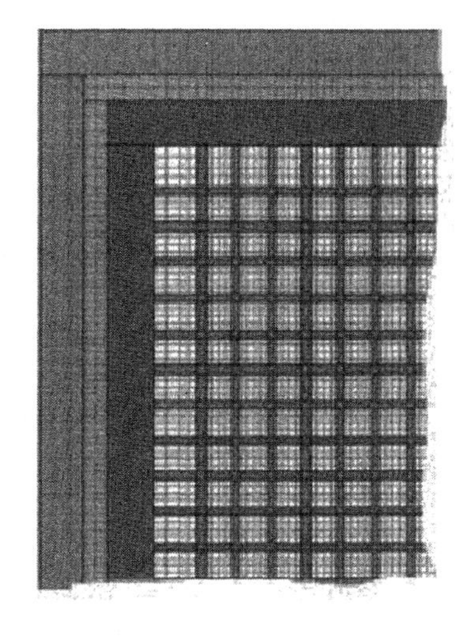

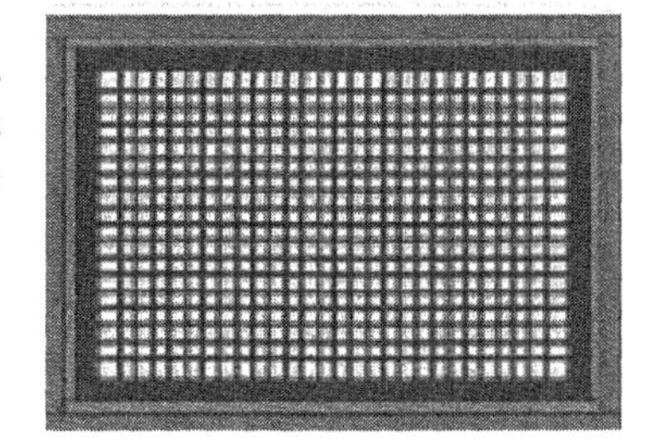

Figure 4. Geometry of Case 1 and Case 2 with typical flow patterns (exhaust mode) and typical cross section (with and without grid displayed) through the checkerwork.

Computational Grid

Using CFD to calculate flow (including heat transfer) in many parallel pipes, although rectangular, has one disadvantage: there is a complex circular interaction chain. For example, when the velocity solver changes velocity, the following consequences will happen:

- Change of velocity induces changes in turbulence.
- Change of turbulent kinetic energy causes changes in the heat transfer from the walls.
- Different heat transfer will change temperature of the fluid.
- Temperature change will result in different fluid density (due to state equation).
- Change of density — and hence gas volume — induces a change of pressure.
- Pressure change causes change of velocity field, and we can return to the beginning of the loop.

These dependencies are, of course, valid in any computation of turbulent flow with heat transfer. However, when we calculate unidirectional flow through simple pipes, CFD seems to be overkill and a simpler approach could be sufficient.

However, in this study, we did use full CFD even for the computation of the flow through the checker pipes. It is important to have accurate approximation of the flow profiles inside the pipes so we need at least a 4 × 4 or 5 × 5 grid cell in each cross section of each pipe (plus one grid for the checker refractory). The resulting grid dimension was 157 × 107 × 69 cells (total ~1.16 million). Part of the grid is displayed in Fig. 4.

Temperatures

As was already mentioned, we started with constant temperature of whole regenerator 30°C. Then we introduced hot exhaust gases. The checkers were heated up (the temperature of the gases leaving the regenerator exceeded 400°C) after 15 h. Figure 5 shows the development of temperatures inside the checkers and in the outlet during the heatup period.

As soon as we had the models heated up, we started the reversal cycle – we blew cold air through the bottom channel for 20 minutes, then we switched to exhaust mode (hot exhaust gases from top) for another, again to firing mode and repeat the procedure until we approach the periodic state.

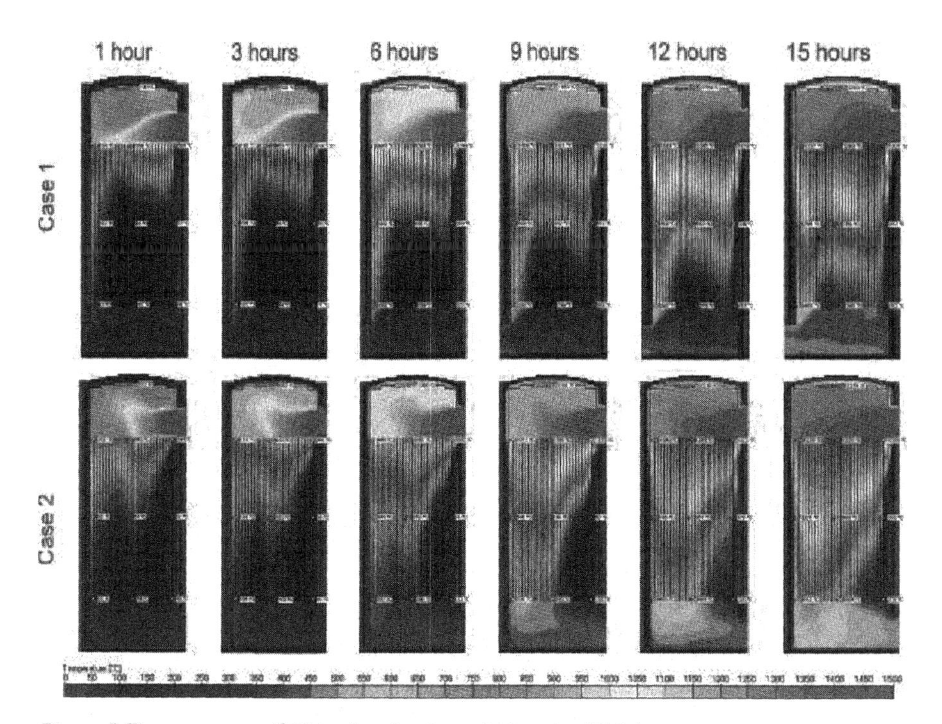

Figure 5. Temperatures (°C) in the checkers during the 15-h heatup.

Effectivity of Preheating

The effectivity of heat regeneration can be calculated as

$$\eta_{reg} = \frac{Q_{exg}}{Q_{air}}$$

where Q_{exg} is the heat contained in the exhaust gases in exhaust mode and Q_{air} is the heat given to the air in the firing mode.

We can see from Table I that Case 2 has a better performance than Case 1, although the maximum temperature in the checkerwork is lower. The reason is that the checkerwork in Case 2 is preheated more uniformly, as can be seen in Fig. 6

Flow Patterns

There is an interesting fact to notice in the flow patterns: the distribution of the air flow among the checkers is far from uniform. Let us concentrate on

Table I. Computation results for Case 1 and Case 2

	Case 1	Case 2
Q_{exg}	6003 kW	6003 kW
Q_{air}	3590 kW	3880 kW
Effectivity	59.8%	64.6%
Preheated air temperature	1090°C	1125°C
Mean checker temperature	791°C	796°C
Maximum checker temperature	1298°C	1281°C

Figure 6. Temperatures of checkerwork and air being preheated in Case 1 (left) and Case 2 (right).

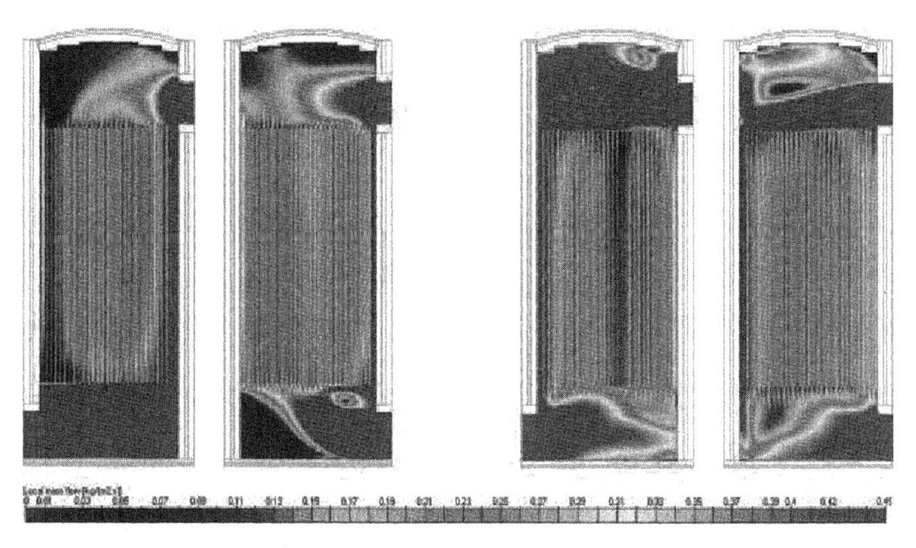

Figure 7. Gas flow [kg/(m^2·s)]: Case 1 firing, Case 2 firing, Case 1 exhaust, and Case 2 exhaust, from left to right.

the firing phase. Cold air enters a regenerator. As can be seen in Fig, 7, especially in Case 1, most of the air flows through the checkers near the port side (on the right side of the picture). What is even worse is that the air in firing stage "prefers" completely different ways than the exhaust gases in the exhaust stage do. This makes the checker temperatures strongly nonuniform and can significantly degrade the effectivity of heat regeneration.

The following are the reasons why the flow, especially in Case 1, is so nonuniform:

1. Because of the momentum of the air entering the regenerator, it does not enter the checkers next to the corner.
2. A relatively big pressure drop starts at the entrance to the burner port above the checkers. Thus, the air prefers flowing through checkers that end close to the burner port because the pressure difference is grater in such pipes.
3. The air that flows through colder pipes (and hence is less preheated) experiences less volume expansion. Thus, it needs less pressure drop than in a hot pipe and a greater mass flow is allowed there. This effect deserves special attention.

The advantage of Case 2 is that effects 1 and 2 mutually eliminate one another. Also, it seems that the flow under the checkers has less freedom and is more stable than in Case 1. That is also why the flow in Case 1 is more complex, more three-dimensional, and less understandable.

The situation in the exhaust mode is similar: the flow in Case 2 is again better distributed and more stable. Only effect 3 works the opposite here: as the gases cool down, they contract and lower the pressure. Thus, in the exhaust mode, the hot gases prefer to flow through colder pipes, which is fortunate.

It is important to realize that the uniformity of the flow strongly influences the uniformity of temperatures, which thus influences use of the heat accumulation capabilities of the checkerwork.

Conclusions

Although there is no steady state in combustion chambers of regenerative furnaces, it is possible to find a time-averaged approximation. However, to model the regenerators themselves, we cannot avoid time-dependent calculation to simulate the process of reversal.

The following are some experiences we noted from calculations of such models.

- The modeling results indicate that effectivity of heat regeneration is very sensitive to the regenerator design. That, on the other hand, means that it may be very beneficial to model the regenerators to find better designs, thus to save much energy.
- In the firing mode (regenerator preheats air), the fact that the air "chooses" an easier way and "prefers" flowing through colder regions of the checkerwork (thus is less preheated) deserves attention. Overcoming this effect could significantly increase the effectivity of preheating.
- Additional models are needed to answer all the questions that are coming from industry. We especially need special models that would predict checkerwork corrosion and plugging. Also, simulation of corroded and/or plugged regenerators would be beneficial, because it would allow us to evaluate the fault tolerance of the device.
- The temperature of the preheated air that enters the burner port is strongly dependent on the vertical coordinate in the port. This should be taken into consideration when setting up boundary conditions for models of combustion chambers.

- We have to simulate (roughly) the process of the furnace heatup and then we have to simulate the reversal process until we reach a periodic state. This makes the simulation very laborious.
- For the checkerworks, at least if they are made out of simple pipes, a full CFD model is overkill and a simplified model would dramatically speed up the calculation without spoiling the accuracy.
- One of the reasons why the model is so time- and memory-consuming is that for each (even rectangular) pipe we need at least 4×4 or 5×5 grid cells per cross section to properly evaluate the complex circular interaction chain of velocity $\rightarrow$ turbulence $\rightarrow$ heat transfer $\rightarrow$ temperature $\rightarrow$ density $\rightarrow$ pressure $\rightarrow$ velocity. However, we could avoid this labor by combining CFD with engineering calculations.
- The simplified model will allow us to calculate the regenerators together with the combustion chamber, also coupled with the glass model.

References

1. P. Schill, A. Franěk, M. Trochta, P. Viktorin, and P. Vlček, "Integrated Glass Furnace Model"; in *Proceedings of 5th International Seminar on Mathematical Simulation in Glass Melting,* Horní Bečva, Czech Republic, 1999
2. B. F. Launder and D. B. Spalding, "The Numerical Computation of Turbulent Flows," *Comp. Meth. Appl. Mech. Eng.,* **3**, 269–289 (1974).

Attenuation and Breakage in the Continuous Glass Fiber Drawing Process

Simon Rekhson, Jim Leonard, and Phillip Sanger
Cleveland State University, Cleveland, Ohio

Work is in progress to define and demonstrate a more efficient glass fiber drawing process using a largely standard production drawing tower. The computer model is used to predict the forming cone shape, forming stress, and break rates for a specified set of process variables. The results of computations for cone shape and forming stress are found to be in agreement with literature data. Calculated break rates are in poor agreement with our data, so the work continues on both the model and the process.

Introduction

This project, in its overall scope, focuses on breakage reduction in continuous glass fiber drawing process. Glass fibers are used to reinforce polymer matrix composites in making plastics for the transportation, marine, and construction industries. The glass fiber industry sells over 1.15 million tons of product annually but also produces a significant amount of unrecyclable waste that is sent to the nation's landfills. The waste is due to fiber breakage during the drawing process. Each time a filament breaks, it takes several minutes to restart the production line. During this time molten glass oozes into the waste dump. Our project aims at the design and demonstration of a more efficient process. It has two parts: (1) process analysis, computer simulation, and process design, and (2) demonstration of a more efficient process on our piloting 200-filament glass fiber drawing unit. This paper focuses largely on several elements of process analysis and computer simulation.

The Glass Fiber Drawing Tower

The unit is an industrial-scale, single-position, dual package winder. The winder is capable of turning from 1200 to 4000 rpm, resulting in final fiber speed of 65–230 kph. The fibers are drawn by a small but industrially representative bushing with 200–400 tips. The bushing is fed by an electrically powered marble melter (Fig. 1). The melter temperature and the bushing temperature are controlled independently of each other. The system has a maximum power capacity of 45 KVA. The system holds 30 lb of molten glass and has maximum melt rate, and thus maximum throughput, of 40 lb/h. The molten glass head is controlled ±0.1 in. using a nuclear level

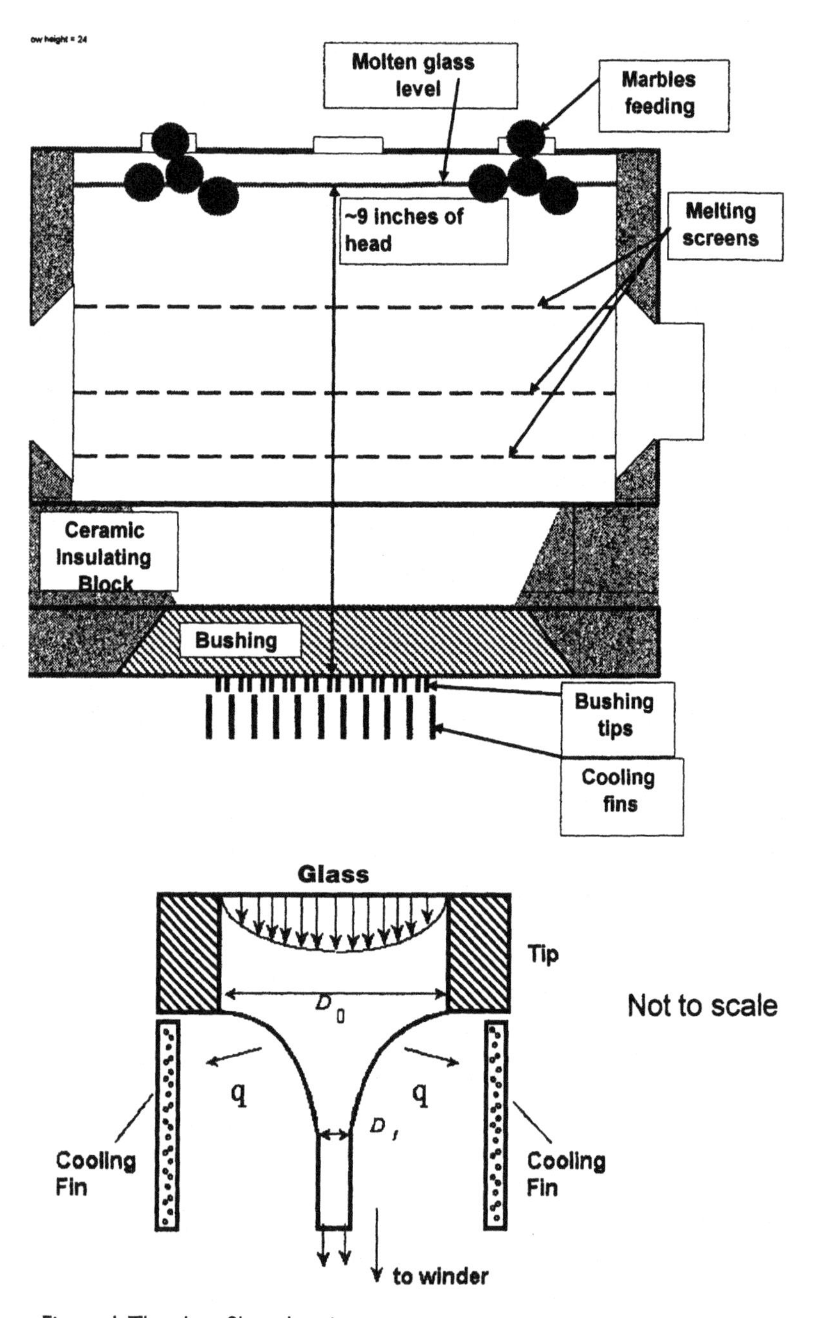

Figure 1. The glass fiber drawing tower.

detector. The system has a digital camera for visualizing the forming cones and an infrared spot pyrometer for measuring the temperature of any location on the bushing.

The Drawing Tower as Instron Testing Machine

In the glass fiber drawing process the molten glass, which is more viscous than sour cream, is oozing through the bore of the tip very slowly, about 5 m/h. One inch below it moves at 180 km/h pulled by a winder. This acceleration stretches the stream of molten glass and reduces its diameter from 1.7 mm down to about 10 μm (which is thinner than a human hair). To quickly visualize such an attenuation, scale the stream diameter up by a factor of approximately 100 and think of a dinner plate collapsing down to the size of pencil lead in a distance of only 2 cm. There must be a tensile or pulling force that imparts such a dramatic shape change. This pulling force effectively tests the glass filament for strength. In 1 h a production machine pulls enough glass filament to wrap the globe almost 10 times. The machine tests 100% of this filament for strength, effectively searching for a weak link. At any given moment the machine tests a short segment of the continuous filament, about 10 mm long, as explained in Fig. 2. One can think of the fiber drawing machine as an Instron testing machine that tests many billions of parts each hour for strength. When one of these "parts" breaks, the broken end of the filament wobbles around and breaks neighboring filaments, the breakage quickly propagating until the line stops.

The concept of breaking a liquid should be a familiar one. Gently pull a small lump of Silly Putty in opposite directions, and it stretches into a filament. Pull it as quickly as you can, and it breaks with sharp edges. Assume the velocity of human hands to be ~10 m/s whereas the winder speed is 50 m/s. The viscosity of Silly Putty is ~50 000 poise, whereas the viscosity of molten glass near the solidification temperature of 900°C is about 20 times higher. Evidently, glass filaments are much stronger.

Order-of-Magnitude Analysis

Now that we expect the jet of molten glass on a rare occasion to break, we need to determine the likely location for such a break. In our analysis of attenuation we actively draw on the classical paper by Glicksman.[1]

Let us start with the upper jet region, which extends approximately over

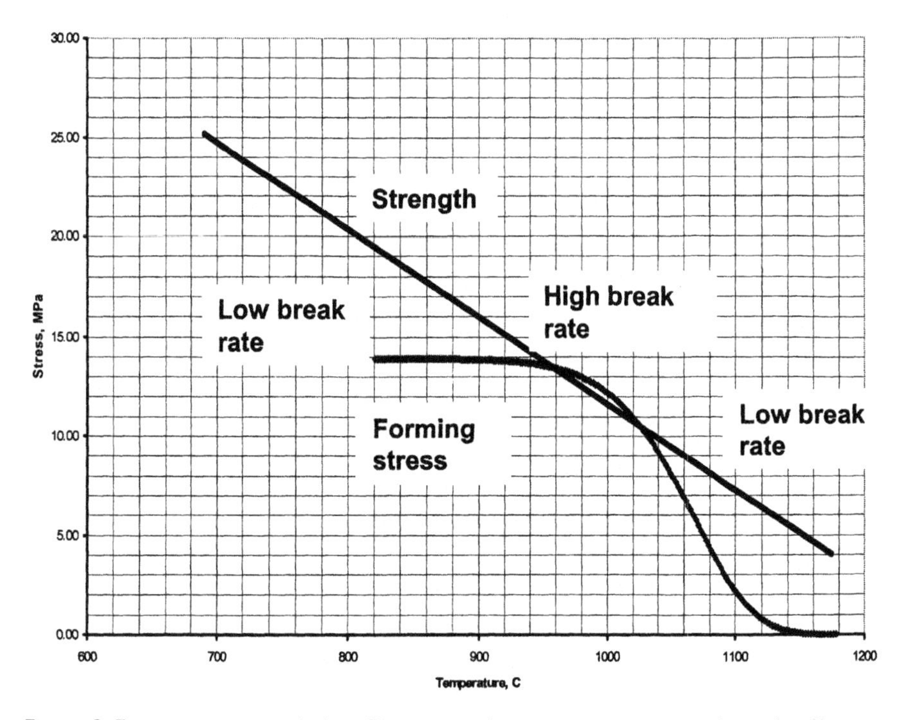

Figure 2. Forming stress and glass fiber strength versus temperature along the filament temperature in drawing. The probability of failure is high within the range of 950–1050·C, which is typically 15–25 mm below the tip.

the first 3 mm (see Figs. 3 and 4 for a shape of the jet). The temperature in this region drops from about 1180°C (2156°F) down to ~1100°C (2012°F), the viscosity increases from ~1000 to ~3000 poise, and the radius reduces from ~900 to ~100 mkm. Thus a jet of the (relatively) low-viscous liquid experiences the attenuation by a factor of approximately 9 in units of radius and 81 in units of area. The stress needed to induce this attenuation is tens to hundreds of kilopascals. As shown in Fig. 2, this stress is much lower

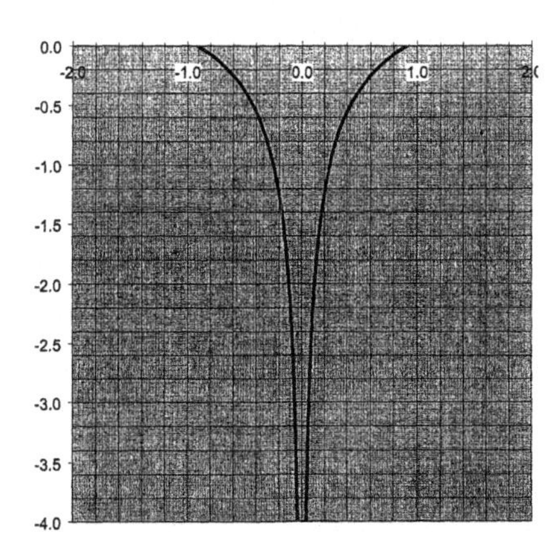

Figure 3. The shape of the cone. Both axes are in millimeters.

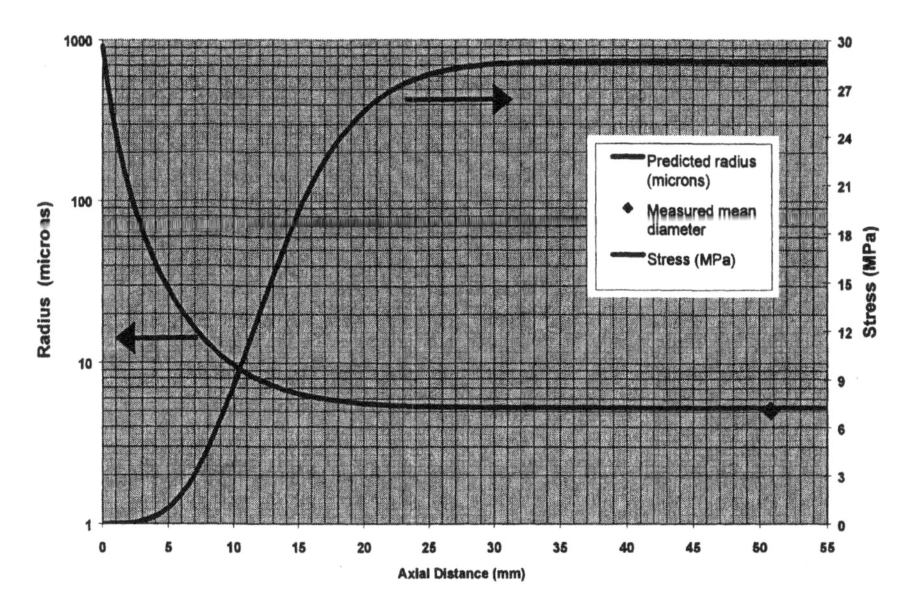

Figure 4. Calculated filament radius and tensile stress versus distance from the tip.

than the strength of the melt and therefore the break level in the upper jet region is unlikely.

The central jet region extends over some 20 mm — see Fig. 4, which clearly shows the jet shape owing to the log (radius) *y*-axis. The temperature in this region further drops from ~1100°C (2012°F) to 900°C (1650°F), the viscosity increases to 1 million poise, and the radius reduces from 100 to ~5 mkm. Here, a jet of the high-viscous liquid experiences the attenuation by a factor of 20 in units of radius and 400 in units of area. Because force is largely constant along the jet, the area reduction of 400 and a viscosity increase by a ~1000 results in a stress increase to 15–30 MPa (see Figs. 2 and 4). Using the Weibull cumulative distribution function with the parameters indicated in Table I, we estimate that the probability of failure in the central jet region is orders of magnitude higher than in the upper jet region. Thus the focus of this work requires accurate evaluation of the jet shape and its temperature in the lower part of the central jet region.

The Forming Cone Model

Glicksman[1] showed that at the central jet region a good fit to attenuation data is achieved using his energy and momentum equations 23 and 24,

Table I. Input variables

Variables	Symbols	Calc's in Figs. 4, 5	Calc's in Figs. 7, 8	Calc'n in Fig. 11	Units
Weibull parameters					
	β	3611-1.68T(°F)		2500	MPa
	m	6.5		6	
Process input variables:					
Tip bore temperature	T_{01}	2175	2100-2260 (Fig. 7)	2050-2300	°F
			2210 (Fig.8)	T_{01}-40	°F
Tip exit temperature	T_{02}	2150	T_{01}-30	T_{01}-40	°F
Winder speed	V_L	1436	1582.2	1150-3065	m/min.
Molten glass depth	H	5	7	5	inch
Process and material constants:					
Number of tips in the bushing	n	1	1	198	
Tip diameter	D_o	0.072	0.067	0.072	inch
Tip length	L_o	0.2	0.2	0.2	inch
Density of glass	ρ	2500	2500	2590	kg/m^3
Nusselt number	Nu	0.42	0.5 (Fig.7)	0.4-0.6	
			0-1.2 (Fig.8)		
Glass heat capacity	C_p	0.31	0.31	0.32	Btu/(lb.°F)
Emissivity of glass	ε	0.3	0.3	0.3	
Ambient air Temperature	T_a	464-1148	68	800	°F

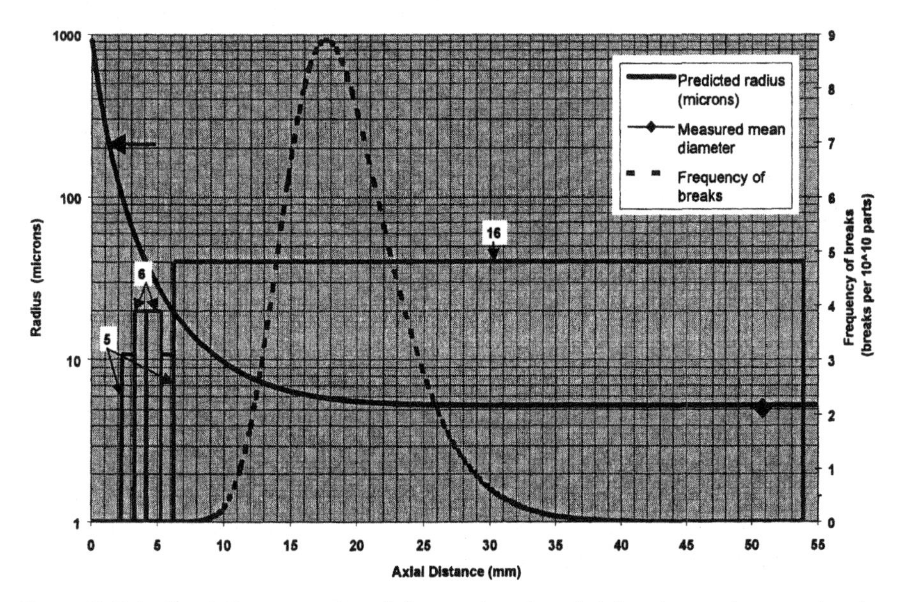

Figure 5. Calculated filament radius (left y-axis) and probability density function breaks (right y-axis) versus axial distance along the filament. The histogram of breaks is data by Koewing and Love.[2]

which are written by neglecting surface tension, gravity, and air drag and by assuming radiation and conduction to be negligible compared to forced convection. This is an important result because it enables us to use closed form solutions for reasoning purposes.

Our calculations presented in Figs. 4, 5, 7, 8, and 11 include surface radiation per Glicksman's equation 13 and neglect bulk radiation. They also neglect air drag because we limit our calculations to 1 in. below the tip.

The input data are given in Table I and, as are the Weibull parameters used for predicting fracture. The results of calculations are shown in Figs. 4 and 5. The location of a maximum in the frequency of breakage is consistent with Koewing and Love's data.[2] The calculated maximum seems much narrower than suggested by the data but this is mostly due to filament diameter changing so slowly with distance that the exact distance from the tip is difficult to determine from diameter data.

Allocation of Responsibilities between Melter and Forming Machine

We will now analyze contributions of the melter and of the forming machine to filament diameter variation and break rates.

Examining Fig. 6 shows the variation of key process variables over the bushing width and the consequences of this variation. As cooling occurs off the walls of the bushing it is logical to assume that the glass melt is cooler at the walls of the bushing and hotter in the middle. Hence the viscosity is lower and, per the Hagen-Poiseuille equation, the flow rate v_0 through the tips is higher in the middle and lower on the periphery. Because the winder speed is the same for all filaments, those filaments, which are in the middle, are thicker than those at the walls. This is how the melter contributes to the filament diameter variation — one of the critical-to-quality characteristics for the end user.

Furthermore, the colder tips at the walls yield shorter (L) and stiffer (η) cones, which, according to the stress equation in Fig. 6, leads to a higher stress and therefore to higher break rates.

In addition, the Hagen-Poiseuille formula reminds us of the melter's responsibility to maintain the hydrostatic head ρgH constant in time. Thus the melter affects both the filament thickness variation and break rate.

We now move on to the key element of the forming machine: the fins. These provide a heat sink for radiative cooling of cones and for cooling of tips if the fins overlap them. Therefore the fins may also affect the viscosity in the tip's bore, just like the melter does. In addition, by shrinking the cone, they shorten the distance L over which the viscosity rises to the solidification viscosity of ~106 poise. Once again, according to the stress equation in Fig. 6, shortening the distance L leads to higher stress and breakage rate.

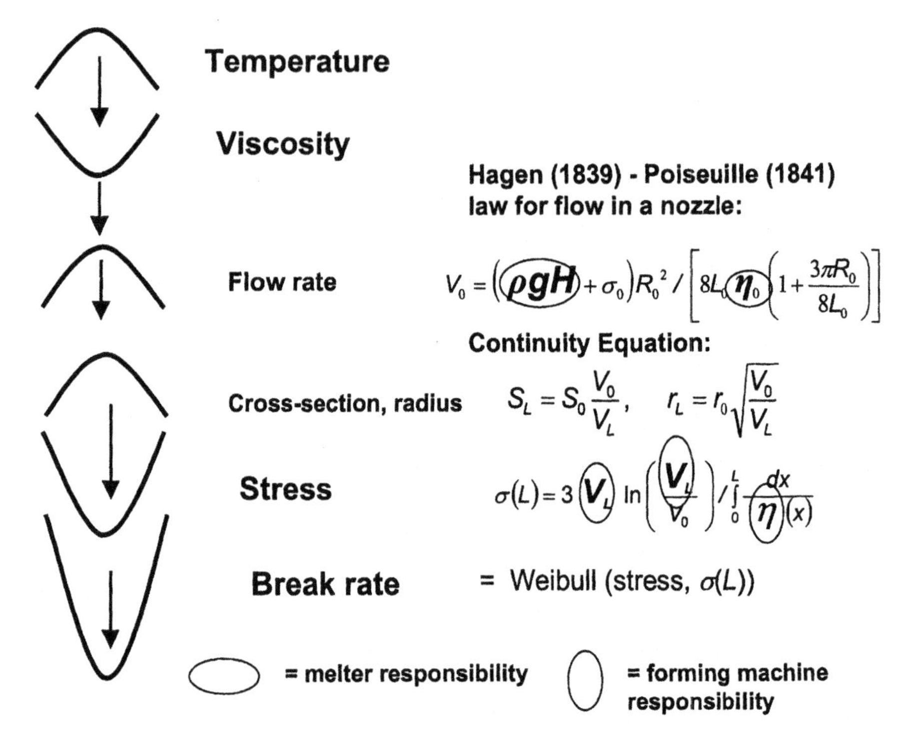

Nomenclature: ρ = density, g = gravity constant, H = molten glass depth; σ_0 = stress from the winder; $R_0 = r_0$, L_0, S_0 = tip bore radius, length and cross-section; r_L, S_L = filament radius and cross-section; V_0 = flow rate in the tip, V_L = winder speed.

Figure 6. Allocation of responsibilities.

Main Effects of Key Process Variables

The analysis above identifies the length and average viscosity of the cone as key product attributes so far as filament breakage is concerned. The closed form solution to the energy equation suggests that the cone length is defined by the ratio of mass flow rate and Nu number. The former is largely controlled by temperature variation in the bushing and the latter by the cone cooling conditions. Figures 7 and 8 show the effects of temperature variation over the bushing and of cone cooling conditions on jet shape and drawing tension. As shown in Fig. 8, our model agrees well with Glicksman's experimental data for the cone shape. The value of forming stress is similarly in agreement at 10–15 MPa.

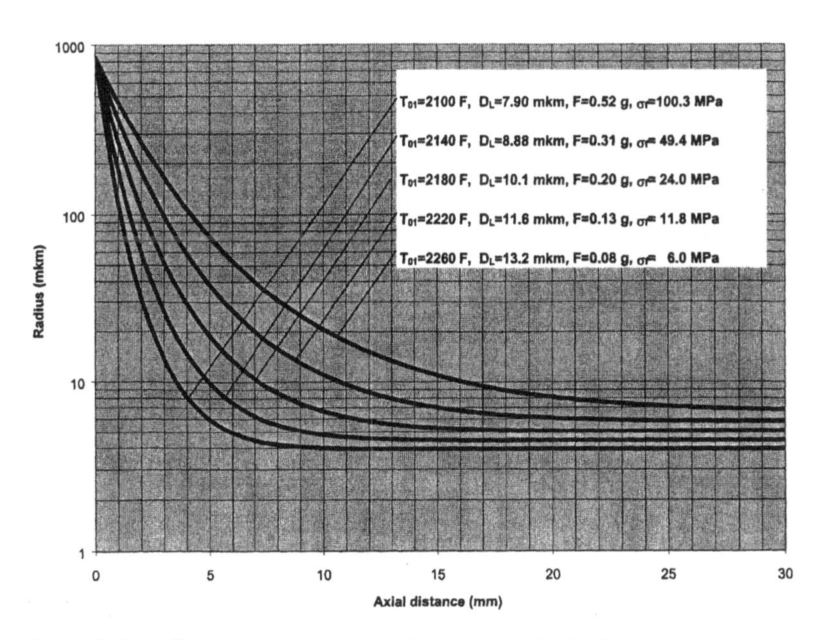

Figure 7. The effect of temperature variation over the bushing on cone length, filament radius, and forming stress.

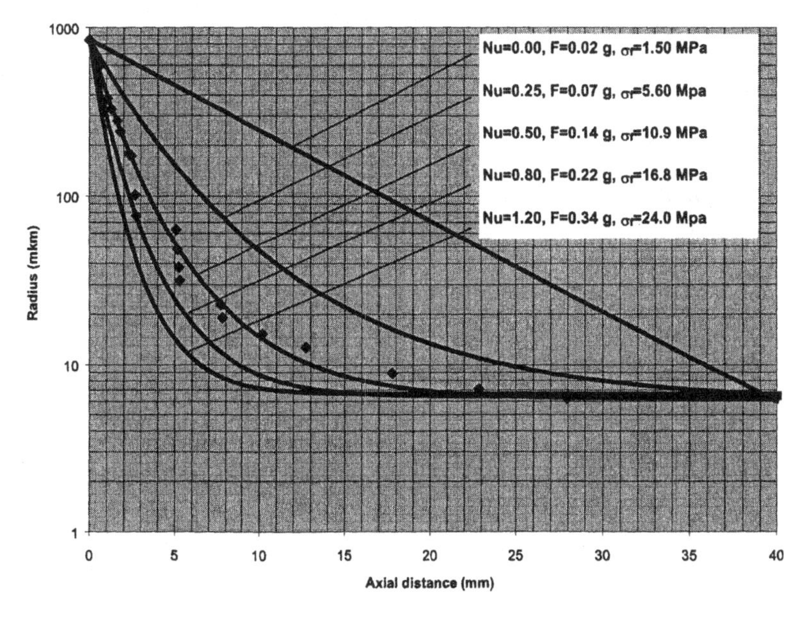

Figure 8. The effect of cooling on cone length and forming stress; diamonds are experimental data from Ref. 1.

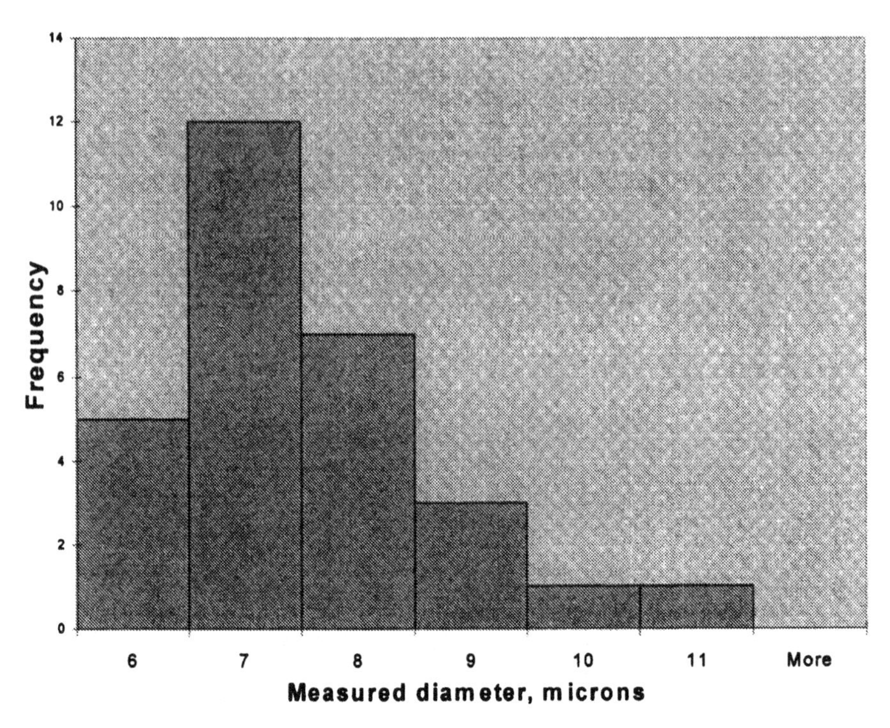

Figure 9. Filament diameter distribution for a sample of 29 filaments.

Validation

A series of designed experiments were run on the first generation system. The goal was to determine the system's performance across the broadest range of operating conditions possible in order to validate the model. The variables in the experiments were the winder speed, the bushing setpoint temperature, and the height of the cooling fins. Their ranges were 1200–3200 rpm (1150–3500 m/min) for the winder speed, 2175–2275°F for the bushing setpoint temperature, and the cooling fins positions from overlapping the tips by 1 mm down to 2 mm below the tips. The responses were runtime to breakage, yarn tension, and filament diameter distribution.

Figures 9–11 show results of 25 validation runs. For each test the filament diameter distributions were measured and characterized by parameters of normal distribution; these were used in calculations of breakage. Figure 9 gives an example of measurements for a set of 29 filaments and Fig. 10, an example of calculations for the entire population of the bushing.

Figure 11 shows run times versus stress for our tests in comparison with computations.

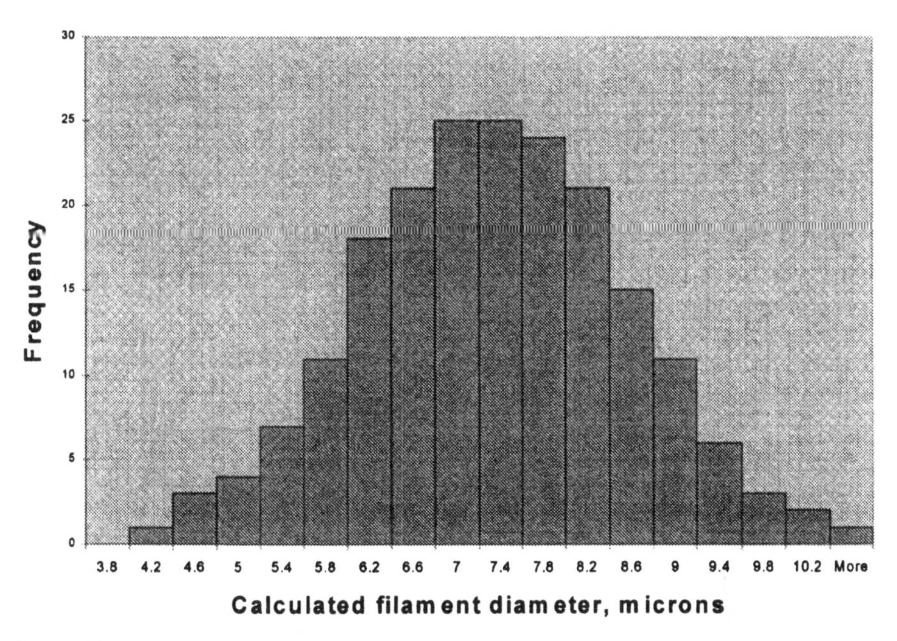

Figure 10. Filament diameter distribution for the entire population of 198 filaments calculated from mean and standard deviation found for a 30-filament sample.

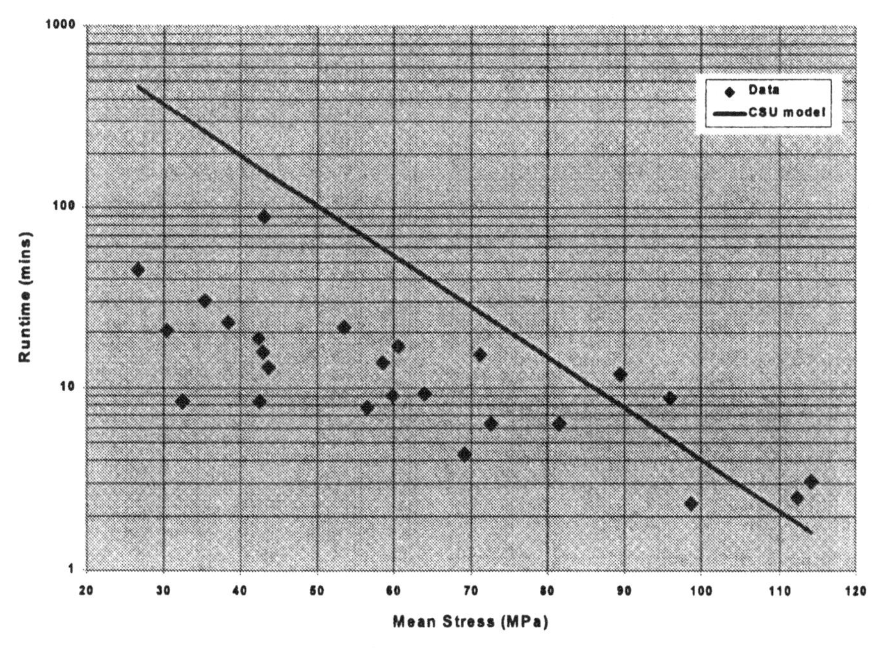

Figure 11. Run time versus mean stress per tests (symbols) and calculations (line).

The best performance we achieved in these tests was 90 min, which is quite poor for a 200-tip drawing tower. This performance partly reflects our learning curve and partly is due to insufficient maximum melting and fining temperatures, short residence time, and inferior quality of 10-year old marbles. The second-generation drawing tower, which has all of these features improved, is currently in operation.

Our computer model predicted the performance up to 0.1 breaks per hour, which is closer to what we are aiming to achieve, but the model is also still in need of critical examination and validation.

Conclusions

Our calculations of attenuation yield the correct cone shape and forming stress values for the central jet region. The probability of breaks and run-time were evaluated using Weibull analysis of brittle fracture. The validation tests were run using the industrially representative bushing with 200 tips and dual package winder. The bushing was fed by an electrically powered marble melter. Work is in progress to improve the performance of both the model and the drawing tower.

Acknowledgements

We are indebted to our colleagues from PPG and Johns Manville for their patient and thorough explanations of glass fiber drawing technology. CSU graduate students Umashankar Sistu and Zhongzhou Chen creatively assisted with theoretical work and computations. The fiber drawing tower was designed and built by PPG using some of their standard production components and a custom-designed marble melter. This project was supported by DOE grant DE-FC07-02ID14347 and by grants from PPG, Johns Manville, U.S. Borax, and Schott Glas.

References

1. L. R. Glicksman, "The Dynamics of a Heated Free Jet of Variable Viscosity Liquid at Low Reynolds Numbers"; pp. 343–354 in *Trans. ASME, J. Basic Eng.,* (1968).
2. J. Koewing and T. Love, PPG Industries, private communication.

Energy Conservation Opportunities in the Glass Industry

John D'Andrea
Southern Company Energy Solutions

Introduction

In the manufacturing community, energy consumption can be one of the major costs of production. However, production and employee issues typically consume the majority of time for managers, and energy is generally an afterthought.

Energy prices have risen considerably in many areas of the country over the last few years and are expected to remain at these levels or increase in the next few years. Now more than ever, appropriate management of energy can yield great financial returns as well as increased production efficiencies. With proper focus, energy costs can be reduced as much as 30–40%. Several major systems, such as compressed air, lighting, steam, and motors, that can yield great energy-saving potential when managed properly.

Compressed Air Systems

Compressed air systems are found in almost every manufacturing plant. These systems are significant energy users in many glass manufacturing operations. They also offer some of the greatest opportunities for energy conservation and production efficiency gains. Upgrading aging and inefficient compressed air systems using a total system approach offers the greatest opportunity for energy savings. By taking a system-level focus, greater optimization can occur and major problems can be corrected. Typical problems include piping bottlenecks, inadequate control systems and storage systems, leakage, and excessive system pressures. Controlling and correcting these problems can save 15–20% of total annual energy expenditures. Typically, project economics show simple paybacks of 18–24 months.

Lighting Systems

Lighting is generally one of the last items considered when looking at managing costs. In the typical manufacturing plant, lighting can be up to 10% of the overall energy expenditure. Optimizing interior lighting systems can significantly reduce energy costs. Typical inefficiencies include the type of lighting and ballasts used and inadequate control systems — these issues can lead to paying more for lighting than is necessary. Proper lighting levels also contribute to production efficiency. Typically, lighting optimization projects have simple paybacks of 18–36 months.

Steam Systems

In many manufacturing environments, steam is a major player in production as well as energy consumption. Fossil fuel boilers are typically used for steam production. In some parts of the country, electric or electrode boilers are now being used in tandem with fossil fuel boilers to produce steam. This is often cost effective when electric tariffs are based on marginal pricing or off-peak consumption. This fuel switching arrangement allows a facility to run an electric boiler when electric prices are low and a fossil fuel boiler when fossil fuel prices are low (gas prices are typically highest in the winter and lowest in the summer, whereas electric prices are typically highest in the summer and lowest in the winter). There are also numerous components of a steam system that if not maintained and managed properly can consume large amounts of energy. Problems with water treatment, steam traps, and inadequate condensate return systems are typical inefficiencies. Although steam systems typically have a high initial cost, existing systems many times can be optimized to yield great savings without having to completely replace boilers or other system components. Typical steam system projects can yield a simple payback of 18–48 months.

Other Energy Conservation Opportunities

The systems mentioned above normally offer the greatest opportunity for energy savings when controlled and managed properly. Many other systems and components in an industrial environment can also offer good energy conservation opportunities. Motors can be a major energy user in most production facilities. Variable speed drives can help to save significant money as well as maintenance costs in many applications. The use of high-efficiency motors is another consideration in many applications. Distributed generation can help save significant money through peak shaving and by help customers take advantage of utility tariffs designed to control system peaks. These systems can also have a positive impact on power quality as well as provide emergency backup power to critical production systems.

Water conservation opportunities also offer many customers the opportunity to save significant amounts of energy.

Conclusion

It makes no difference how old or new your facility is, there are always opportunities for energy conservation. Many of these decisions take a back seat to day-to-day production concerns but can truly help in improving production while lowering overall production costs. There are numerous places you can turn to for help. Most local utilities have services that help customers manage their energy costs. Regional and national energy services companies can help you determine savings opportunities as well as manage these projects for you. The Department of Energy and other governmental groups can also be a great source of information and assistance in helping you effectively manage your energy costs.

Process Control

Application of Fast Dynamic Process Simulation to Support Glass Furnace Operation

Olaf Op den Camp and Oscar Verheijen
TNO Glass Group, Eindhoven, The Netherlands

Sven-Roger Kahl
REXAM Glas, Dongen, The Netherlands

TNO Glass Group has developed a fast glass process simulator (GPS) that can easily be used by operators and glass technologists to obtain important (and previously unavailable) information about the impact of furnace operation on furnace performance. Not only does GPS provide online information on flow, temperature, and redox distribution of the total glass melt, but by using GPS the user can also check on the impact of changes in furnace settings or disturbances on important quality indicators such as color, sulfate retention, residence time distribution, and fining quality. After an introduction to the simulation of the flow, temperature distribution, and redox state of the glass melt and the impact of changes in process settings on the fining performance and on the color of the glass, the architecture of GPS will be briefly explained. The paper is concluded with a discussion on the pilot GPS that is being applied in an emerald green container glass-producing furnace at Rexam Glass Dongen in the Netherlands.

Introduction

In the battle for more flexible glass production, lower energy consumption, reduction of emissions, and better quality, it is not always easy to select the best operating conditions and to keep a glass furnace under tight control. Therefore, in the last decade, automatic control of furnace and forehearth temperatures has become very popular. All kinds of control systems are on the market, ranging from conservative single PID control to sophisticated model-based predictive control. Most of these systems simply control a single temperature at the inside of the crown or in the glass near the bottom. Other, more advanced systems, specially designed for forehearths, control the total glass temperature curve just before the exit. However, temperature is only one of the key parameters in glass production. Glass quality depends on the total temperature and redox history of the glass, which indicates that, in addition to temperature, the total flow and redox state distribution of the glass melt are of main importance. At the same time, energy consumption, cullet percentage, emissions, and so on are severe constraints on operation.

To support operators and glass technologists in the selection of (improved) furnace settings, TNO has developed a furnace navigation system called the glass process simulator (GPS). GPS is a fast dynamic simulator of all processes in a glass furnace: melting, fining, mixing, and homogenizing. The so-called GPS monitor is used as a soft sensor to provide online information with respect to glass flow, temperature distribution, emissions, energy consumption, residence time distribution, melting/fining performance, redox state, and glass color. Important operating conditions, such as hot spot temperature and position, spring zone location, minimum residence time, and expected glass color, are automatically determined. Also, additional glass process quality indicators are determined so the operator can easily judge the current and expected future performance of the furnace with respect to melting, fining, and homogenizing.

The online GPS monitor is coupled to the furnace distributed control system (DCS) in order to enable data transport from the DCS to GPS. In this way, all actual process settings (pull, pull distribution, total amount of gas, gas distribution, air flow, etc.) are transferred to GPS. Next to these actual process data, glass quality data such as seed count, color, and redox state and glass batch composition data are entered in the online GPS each moment these data are available from laboratory measurements. The combination of process data, soft-sensor data, and data on glass quality enables the identification of the most important process parameter(s) for excellent glass quality.

Within GPS, the user can easily make a copy of the current furnace state and use an offline GPS predictor as starting point for investigations into the impact on furnace performance of intended changes in process settings and unintended process disturbances. The user can easily insert these changes through a dedicated user interface. The offline GPS will determine the response of the furnace to these changes as function of time. Within hours, the GPS predictor predicts furnace performance over a period of several days. In this way, several scenarios of possible improvements in process settings can be tested without influencing the real process and consequently without production losses due to downtime or increased glass rejects. On the basis of the furnace performance information that GPS provides, the user can decide on the best possible process settings for the furnace. Because GPS determines time transient results, even the strategy of changing to different settings can be tested (for example, for color changes) and the furnace can be prepared gradually for a new state. In this way, the user can navigate the furnace into its best performance using GPS.

Because GPS behaves like a real furnace, it can be used as a training tool for new operators. For training, an offline GPS will select at random or on the basis of input by a trainer a realistic time-dependent disturbance (or even a combination of disturbances) to the glass melting process and the trainee will be asked to react to the furnace behavior by adapting the furnace settings. Disturbances might consist of changes in redox state of the cullet, changes in the temperature of the combustion and cooling air (day/night rhythm), a change in the caloric value of the fuel, a change in glass melt properties, or rapid aging of one particular thermocouple.

Furnace Simulation

Together with major glass producers, TNO has developed a detailed model for the simulation of processes in glass melting furnaces: melting of the batch of raw materials and cullet, fining of the glass melt to reduce the number of blisters and amount of dissolved gases in the glass, and homogenization of the glass melt before the glass melt is delivered to the forming section(s). This detailed glass tank model (GTM) is able to simulate any type of furnace with any kind of heating system, such as cross-fired, U-flame regenerative, recuperative, oxy-fuel, gas and/or oil combustion, all-electric, or electric boosting. Moreover, different types of batch chargers (blanket chargers, screw chargers) and different types of mixing systems (such as bubblers and stirrers) are available.

In many projects, GTM has been used to simulate steady-state situations especially to assess the impact of changes in design and time-transient phenomena such as load changes and color changes. Once the temperature and flow distribution have been computed, it is only a small step to determine the residence time distribution and the efficiency of the fining process by means of a particle trace and a redox module. The models have been thoroughly validated against measurements of temperature distribution (probing using a water-cooled thermocouple lance through designated holes in the crown) and of the pressure of the physically dissolved oxygen in the glass using pO_2 sensors. Because in this paper the applicability of GPS in industry is shown for an emerald green glass furnace, a more detailed description is given of the simulation of redox and color as these properties are of key importance for good glass quality.

In glass forming batches, fining agents are added in order to release gases at elevated temperatures (>1350–1400°C). The released gases (e.g., oxygen and sulfur dioxide) oversaturate the glass melt at high temperature, which leads to O_2/SO_2 blisters in the glass melt. Because the partial pressure of the physically dissolved gases in the glass melt (such as CO_2 and N_2) is high compared to the partial pressure of these gases in the blisters that result from fining, these physically dissolved gases will diffuse into the fining blisters. In this way, the fining blisters strip physically dissolved gases out of the glass melt. The fining blisters will grow and, due to buoyancy forces, move toward the glass melt surface to escape from the glass melt. As the fining agents reabsorb the fining gases at decreasing temperature, the fining blisters that have not reached the glass melt surface will dissolve easily at decreasing temperature, which is known as secondary fining. In this way, secondary fining will prevent the occurrence of blisters in the final product, because it lowers the total amount of gases dissolved in the glass melt.

The release of fining gases due to the thermal decomposition of sulfate in emerald green glass and therewith the fining performance of the glass furnace is dependent not only on temperature but also on the redox state of the emerald green glass just after batch melting has been completed. Figure 1 shows the production of fining gases as function of temperature for different redox states (expressed by the $Fe^{2+}:Fe_{tot}$ ratio) of the freshly formed homogeneous melt phase after batch melting. It is seen that gas production is very temperature dependent and that the gas production onset temperature shifts to higher temperatures in case of a more reduced (higher value for $Fe^{2+}:Fe_{tot}$ ratio) freshly formed glass melt. In GPS, the fining performance of the glass furnace is quantified by the calculation of the so-called fining index (FID). This quality parameter takes the effect of glass melt temperatures and redox state on the fining process in the glass melt into account. A high value for the fining index represents a well-fined glass melt, whereas a low value for the fining index may lead to the presence of seeds in the glass containers.

Other redox components in the glass melt, such as Fe_2O_3 and Cr_2O_3, that are used to color the glass, interact with the fining agent(s). Different valence states of these coloring oxides lead to different colors. Consequently, the color of the glass product not only is a result of the concentration of coloring agents, but also depends very much on the redox state of the glass. Because the production of fining gases and the coloring of the final glass container depend on the redox state of the freshly formed glass melt after batch melting has been completed, an inline batch redox sensor is posi-

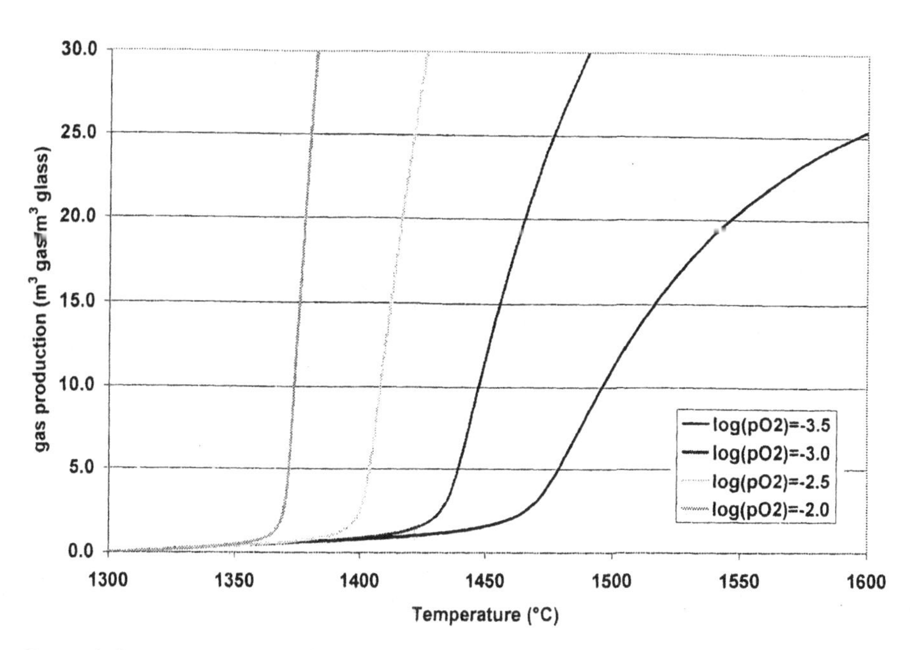

Figure 1. Fining gas production in emerald green container glass as function of temperature and redox state of the initial formed glass melt after batch melting.

tioned in the doghouse area of the emerald green container glass furnace, close to the batch blanket (Fig. 2). This redox sensor provides a continuous mV signal to GPS that represents the local redox state of the freshly molten glass. This redox signal enables GPS to predict the redox state throughout the glass-melting furnace and the properties of the final product such as color characteristics, redox state, and sulfate retention. Because GPS is much faster than real time, the effects on product properties of redox disturbances at the entrance of the glass furnace measured by the inline redox sensor can be predicted within a short time.

The coloring components in the glass melt also influence the thermal conductivity of the glass melt. For this reason, coloring agents have a huge impact on convective currents in the melt. The practical observation that glass melt temperatures in emerald green container glass furnaces are very much dependent on redox state of the used cullet confirms this and stresses the need for a quantitative redox and gas distribution model combined with a temperature- and redox state–dependent thermal conductivity of the glass melt and an inline batch redox sensor to provide the initial redox state of the glass melt.

Figure 2. An inline batch redox sensor located in the doghouse area of the emerald green container glass furnace that provides the redox state of the freshly formed glass melt.

The redox model predicts the interaction between redox species, and the effect on color, thermal conductivity, and consequently flow and temperature distribution in the glass melt. On the basis of the concentration of the redox components (fining agents, coloring agents) that enter the batch, the initial redox state of the glass melt at the interface with the batch (as measured by the inline batch redox sensor), and the concentration of dissolved gases in the glass just underneath the batch, the redox model determines the shift in the valence state of the redox species dependent on the flow and the temperature distribution throughout the whole furnace. Also the release of fining gases and the consequent stripping of the physically dissolved gases is simulated. The solution of all simultaneous redox reactions in the glass melt results in concentration distributions of all polyvalent species throughout the glass furnace. Subsequently, these results are used to determine the local glass melt properties such as the heat transfer coefficient, the final glass color, the redox state of the glass product, and sulfate retention. Moreover, the redox model is used to determine the performance indicator for the fining process FID.

GPS Architecture

Although indispensable for the evaluation of changes in design, GTM does not meet requirements with respect to computational speed for the continuous simulation of a furnace under changing process conditions. GPS is however a fast dynamic simulator for a furnace that is based on the detailed

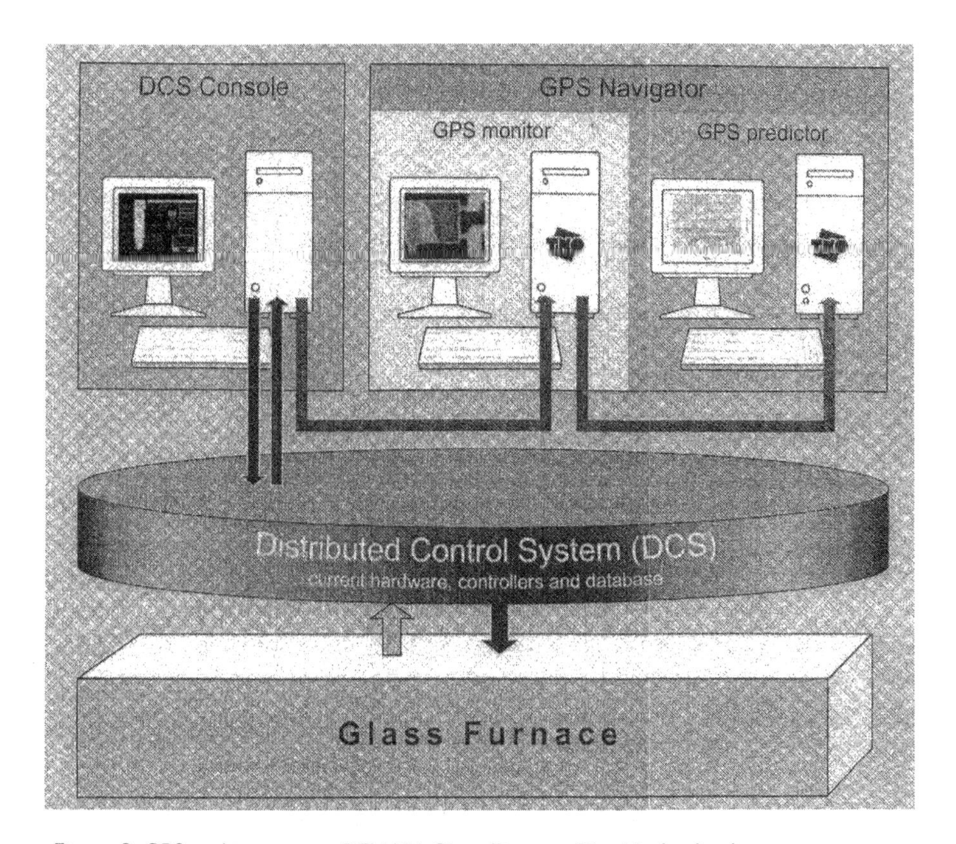

Figure 3. GPS architecture at REXAM Glass Dongen, The Netherlands.

GTM model. A collection of advanced mathematical techniques is used to speed up the model. The resulting GPS is fully based on first principles and is a furnace-specific product.

The online version of GPS is used to monitor the current process. This means that the model runs parallel to the process, using the same settings as the furnace. These settings (pull, pull distribution, gas consumption per burner, direction of firing, and applied electrical power on each of the transformers) are read from the furnace's DCS system. In the application of GPS at Rexam Glass Dongen, GPS communicates with a Honeywell DCS console called a GUS station (Fig. 3). Some settings, such as the pull, are not necessarily available from the DCS. GPS determines the pull in that case on the basis of the gob weight and the cutting speed for each line. Each minute, the furnace settings are read, and based on these settings GPS will perform a (time-transient) prediction of the furnace state for the next 5 min.

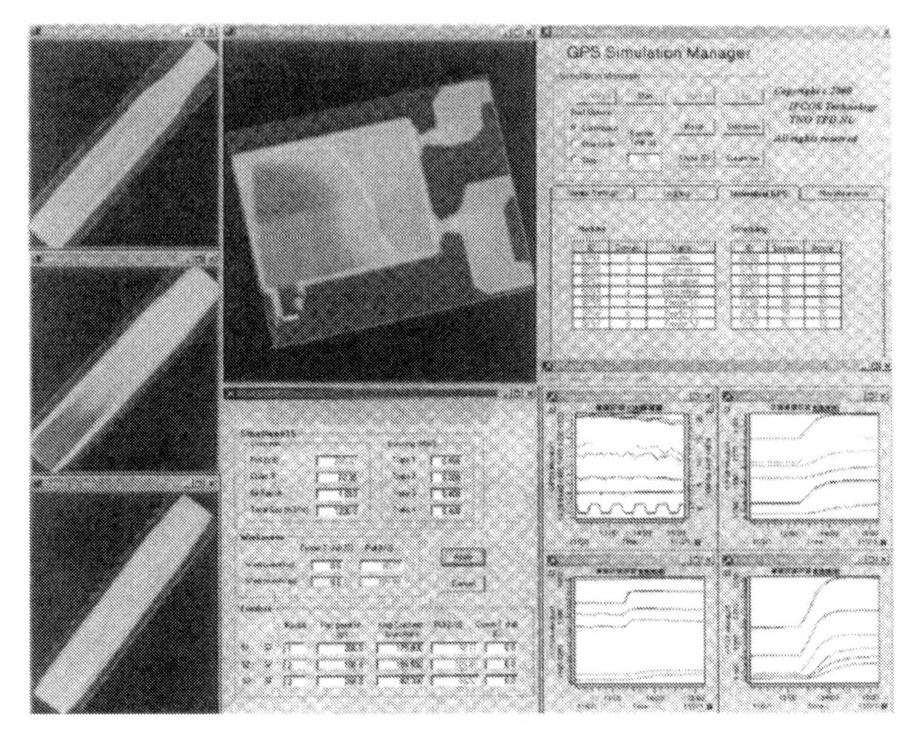

Figure 4. Overview of the graphical user interface of GPS.

Because the simulator is faster than the real process, a model-tracking algorithm is used to force the model to run in pace with the furnace. The model tracker also ensures that the model fits the real behavior of the furnace very well. Model tracking is essential because the user will base decisions on the outcome of the GPS model.

While the simulator is running, the user can study the furnace state variables via a very simple post-processor, in the form of tables, graphs, and figures (Fig. 4). In such a way, the user has continuous access to information with respect to the furnace that cannot be measured by sensors, such as residence time distribution including minimum residence time (for each line), fining performance indicator, hot spot position and temperature, redox state ($[Fe^{2+}]/[Fe]_{tot}$), or produced product color. By means of graphs, the historic trends of these variables are also available. At the same time, a three-dimensional overview of the flow and temperature profile in the entire furnace is given.

By means of this information and the information from the sensors (thermocouples, redox sensors), the user can judge the overall performance of the furnace and, when necessary, can make changes to process settings at an early stage to control the furnace.

To determine the impact of changes in furnace settings before actual application at the furnace, the offline GPS predictor is used. The initial state of the furnace is read by this version of GPS from the GPS monitor (which provides the best estimate of the current furnace state). All settings can be changed in a straightforward user interface. The GPS predictor will immediately take these changes into account and compute the resulting change in furnace state and furnace performance indicators over time. In this way, the impact of changes in furnace settings can easily be tested without any risk of furnace or production problems. The GPS predictor can be used to search for the most optimal settings within a limited time span.

Application of GPS at an Emerald Green Container Glass Furnace

The application of GPS is demonstrated for an emerald green container glass furnace for which a lower energy consumption level is achieved by applying a modified furnace setting in combination with a changed redox state of the glass-forming batch. The task of the described strategy is to keep the color of the produced glass containers within the customer-imposed specifications while maintaining a good fining process so as to avoid glass rejects due to excessive seed count and to run the furnace at a lower energy consumption level. For the emerald green container glass, the main important product (color) specification is the dominant wavelength (DWL) of the produced container. The fining performance of the glass furnace is judged by the fining quality parameter FID.

To decrease the energy consumption level at a constant furnace pull, the boosting power is reduced while a constant combustion energy supply is maintained. The reduction in boosting energy results in a decrease in glass melt temperatures of about 15°C as indicated by a bottom thermocouple (Fig. 5). As a result of the decreasing glass melt temperatures the fining performance (characterized by the FID) reduces, which may lead to the presence of (additional) seeds in the glass product. The predicted DWL of the glass containers barely changes, indicating that the imposed change of glass melt temperature has only a slight impact on the color characteristics of the produced glass (Fig. 6).

To avoid decreased fining performance while maintaining the lower glass

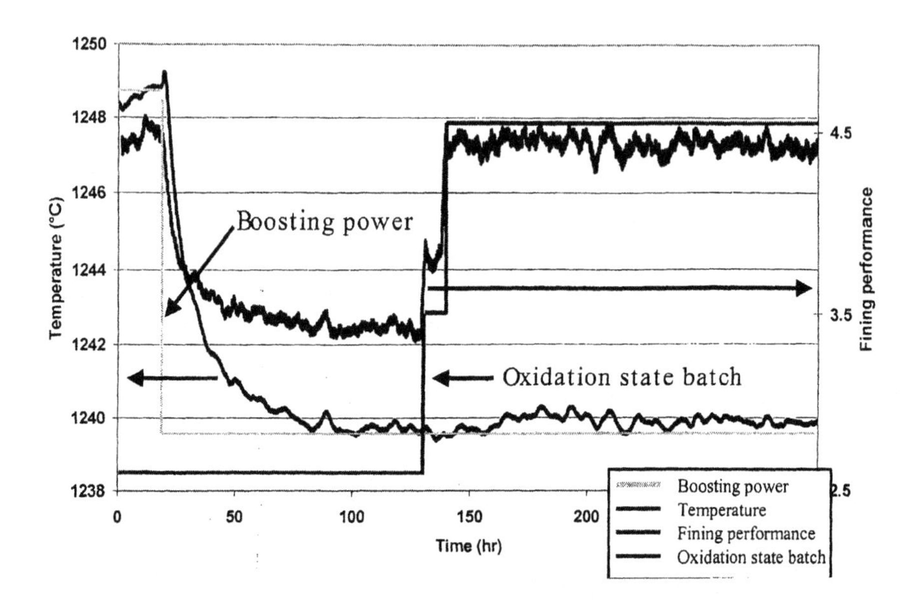

Figure 5. Imposed changes in boosting power and glass batch redox state, and the resulting effect on glass melt temperatures and fining performance as determined with GPS.

melt temperatures (and therewith the lowered specific energy consumption of the glass-producing furnace), the redox state of the glass batch has been changed. As was shown in Fig. 1, the onset for the production of fining gases shifts to lower temperatures with a more oxidized green glass melt. Here, the decreased fining performance of the glass-producing furnace is compensated for by using a slightly more oxidized glass melt just below the batch blanket.

As a result of the change in redox state of the glass batch, the fining performance improves, which indicates that the number of seeds decreases. Although the DWL increases, the increase in DWL is small enough to prevent exceeding the imposed limits on DWL for this emerald green container glass furnace.

Concluding Remarks

This paper described the basics and architecture of the fast glass process simulator developed by TNO Glass Group. GPS is used both as an online and offline tool to supply operators and glass technologists with the maxi-

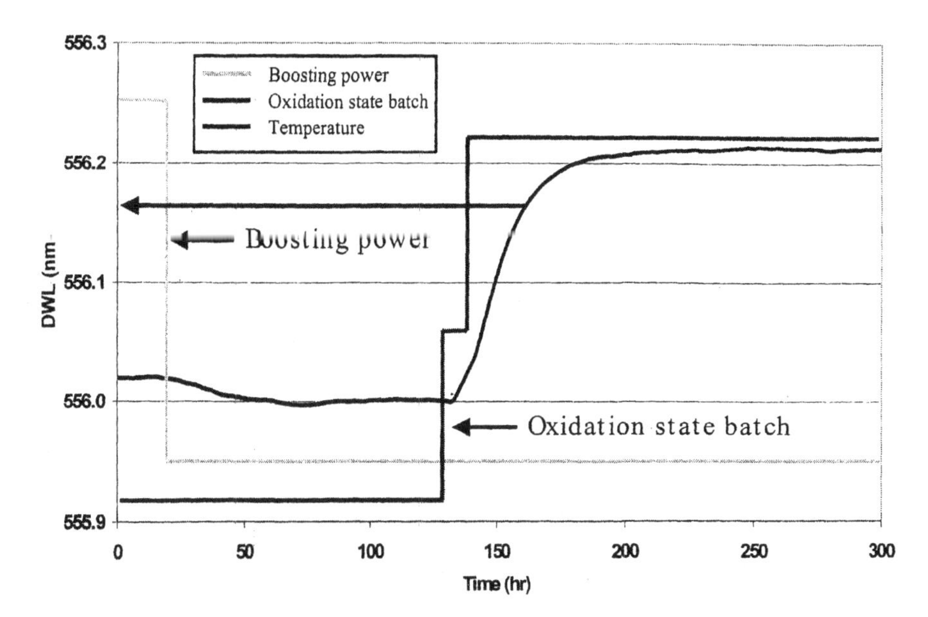

Figure 6. Imposed changes in boosting power and glass batch redox state and the resulting effect on dominant wavelength as determined with GPS.

mum amount of relevant information while operating the furnace. The practical use of GPS has been demonstrated for an emerald green container glass furnace. For this furnace, modified furnace settings and glass batch redox state were determined such that a lower specific energy consumption for the furnace was obtained while maintaining a good fining process and still meeting the imposed color specifications.

In addition to the determination of improved process settings, GPS alone or in combination with inline redox sensors can be used for improved control and stabilization of furnace operation. Especially for green container glass to which large amounts of foreign mixed cullets are added as raw material, control of redox state and early stage compensation for redox variations in the cullet measured by the inline batch redox sensor has become possible.

Acknowledgments

The authors wish to thank Bart Smits and Andries Habraken of TNO Glass Group for their contribution to the work discussed in this paper. Moreover, EET (Economy, Ecology, and Technology), a program of the Dutch government, is gratefully acknowledged for supporting the development of GPS.

Application of Batch Blanket Monitoring System in Glass Furnaces

Jolanda Schagen, Ruud Beerkens, and AnneJans Faber
TNO Glass Group, Eindhoven, The Netherlands

Peter Hemmann and Gunnar Hemmann
Software & Technologie Glas GmbH Cottbus, Cottbus, Germany

Melting processes and glass quality can be optimized by using information on the batch blanket pattern for controlling the batch coverage in industrial glass tanks. For this aim, a batch blanket monitoring system that continuously analyzes batch blanket patterns in industrial glass tanks has been tested by TNO in cooperation with Software & Technologie Glas GmbH Cottbus. Correlations between batch blanket parameters (length, size, covered surface area, symmetry) and process parameters, such as crown temperatures, batch-charging settings, and input of fuel, were demonstrated. These correlations can be used to improve the melting processes.

Introduction

Monitoring batch blanket patterns in a glass tank yields important information for optimization of the melting processes and glass quality. The application of a batch monitoring system based on digitalized video images for monitoring batch patterns in glass tanks has been studied.

The added value of a batch monitoring system is that the batch blanket parameters derived by the batch monitoring system can be used in a control system to control batch chargers and input of fuel to achieve a stable batch blanket and to improve glass quality, when correlations between batch blanket parameters and process and glass quality parameters are determined.

Procedure

The batch monitoring system consists of both hardware and software. The hardware consists of a water-cooled furnace camera and a computer with a built-in frame grabber. The movies recorded by the camera are continuously transferred into digital images. Every few seconds or minutes a snapshot can be taken from these digital images that can be analyzed by the computer program. From such images, information on the current location and coverage of the glass surface by a batch blanket can be derived. Figure 1 shows the schematic overview of the batch monitoring system.

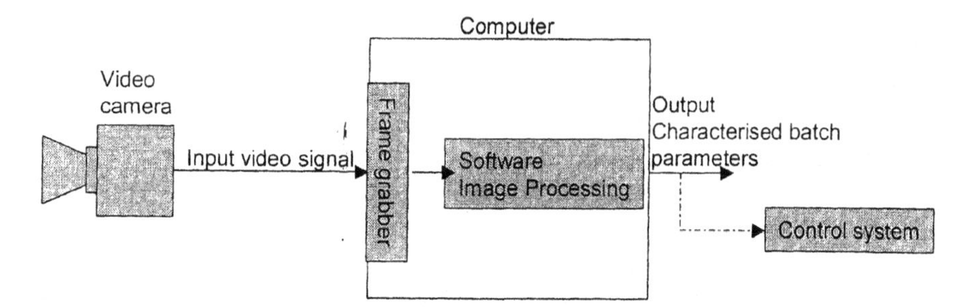

Figure 1. Schematic overview of the batch monitoring system.

The frame grabber converts the analog video signal to a digital format. After this conversion, image-processing software is used to analyze the images. This software contains the following main functions:

- Presentation and sampling of the live video images. Snapshots can be taken on a second scale and the software is able to derive all essential information from the snapshots in less than 1 s.
- Batch blanket recognition by the identification of the area covered with batch blanket.
- Correction for perspective distortion of the camera's view.
- Calculation of previously defined batch parameters (such as batch length, surface area covered by batch), which can be used in current and future control systems.

Figure 2 shows an example of the user interface of the batch monitoring system. The user interface consists of four quadrants. The upper left quadrant shows the real-time images. The upper right quadrant shows the sampled image. The sample frequency is adjustable from several images in a second up to sample times of several seconds. In each sampled image the positions of the surfaces covered by batch can be determined. The identified batch blanket of each sampled image is shown in the lower right quadrant. From this image, batch blanket parameters, such as percentage batch covering the melt surface and batch length, are derived to obtain quantitative data on the shape and dimensions of the batch blanket. The quantitative batch blanket information is monitored in the lower left quadrant and is directly saved in a text file and/or database. The way of demonstrating the batch blanket parameters in the lower left quadrant is defined and can be chosen by the user (process technologists in the glass factory). This can be, for example, in the form of time graphs, average values over an adjustable time, and so forth. The sampled images can be saved if desired.

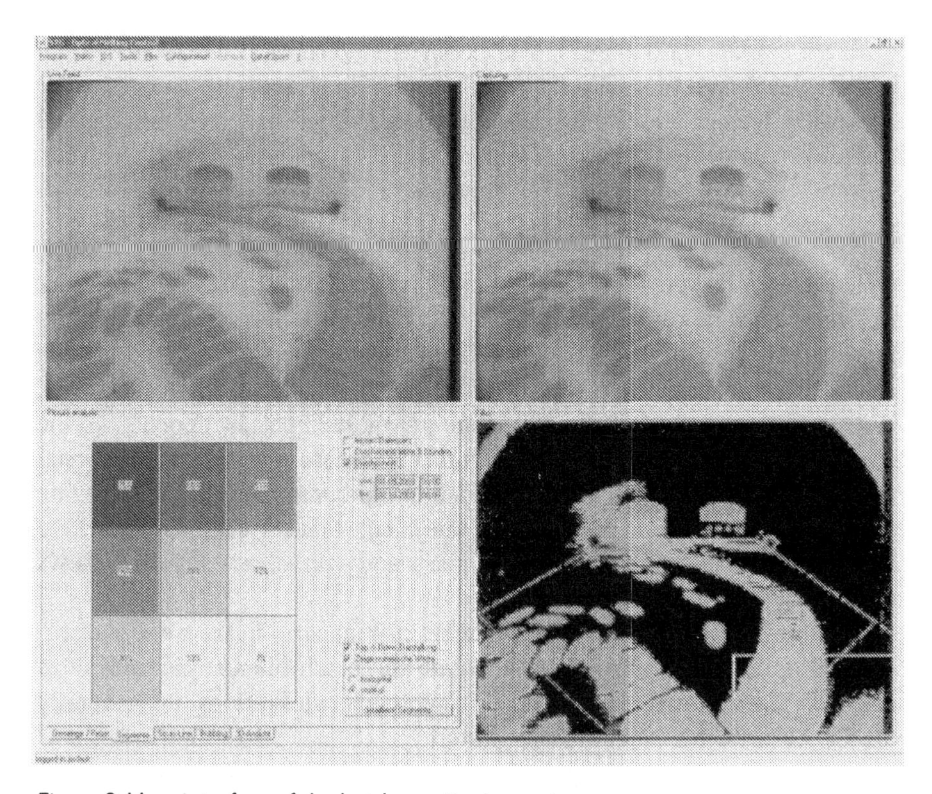

Figure 2. User interface of the batch monitoring system.

Analyzing Batch Blanket Patterns

Different functions are used to achieve the quantitative batch blanket parameters from the monitored images. When the furnace is continuously oxygen fired, analysis of the sampled images can be carried out continuously. However, when the furnace is air fired, reliable batch recognition is possible only during the firing-side change period, when there are no flames. The flames of air-fired furnaces are usually less transparent in contrast to oxy-fired furnaces, and may hinder the view of the batch blanket pattern.

The presence of flames can be derived from the images of the furnace. Only during the change of the firing side, when no flames are detected, are images recorded for batch blanket pattern analysis.

An adjustable region of interest (selection of the most important area in the image) is used to record information about the batch blanket in that area. The identified batch blanket areas in this region will be used to analyze the batch blanket pattern. A correction in the calculation of the dimensions is applied to take into account the perspective distortion caused by the different camera view angles. The perspective distortion means that batch islands positioned far away from the furnace camera appear to be smaller than batch islands close to the furnace camera. After this correction the real surface coverage by batch is calculated.

Batch Blanket Parameters

Quantitative characteristic batch blanket parameters are identified in order to find a correlation between the batch blanket pattern and process and glass quality parameters. The batch blanket parameters are defined in such a way that they result in a representative batch blanket pattern description. Which of the batch blanket parameters are most important depends on the specific furnace where the batch monitoring system is installed. Additional batch blanket parameters can be defined and programmed for each specific furnace optionally. For definition of the batch blanket characteristics, the furnace is divided into small strips in the x and y directions. Figure 3 shows the subdivision of the furnace.

The surface coverage by batch blanket in each strip in the x and y directions is derived and is used to calculate the batch blanket parameters of interest. The following batch blanket parameter definitions are generally used to describe the batch blanket pattern:

- Maximum relative batch blanket length as a percentage of furnace length: Maximum batch blanket length in the x direction in the furnace relative to the furnace dimensions.
- Coverage: Total level of coverage of melt surface by batch blanket in the furnace in m^2.
- Symmetry: Symmetry between the batch blanket surface coverage at the left and right side of the furnace.

$$\text{Symmetry} = \frac{A_L}{A_L + A_R}$$

  where A_L is the batch blanket surface coverage at the left side of the furnace in m^2 and A_R is the batch blanket surface coverage at the right side of the furnace in m^2.

- X_{mean} in m^2: Average batch blanket coverage over the ten strips in the x direction.
- X_{center} in meters: Mathematical center of the batch blanket coverage in the x direction (x component of the center of mass).

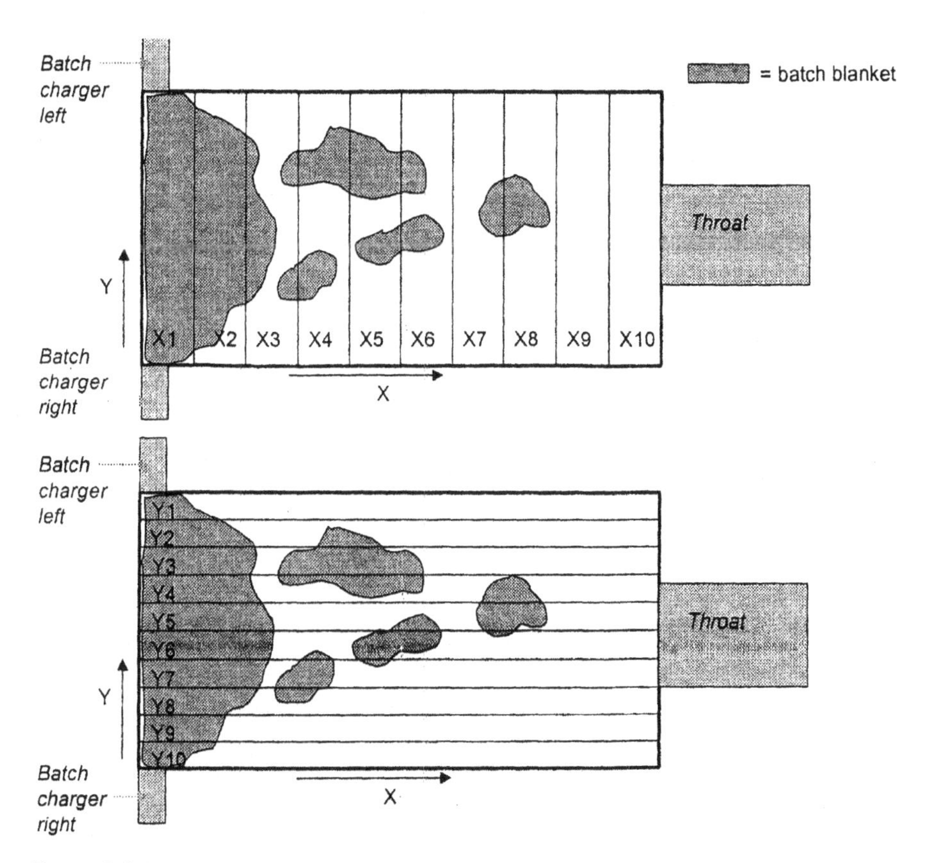

Figure 3. Schematic overview of the subdivision of the furnace into strips.

$$X_{center} = \frac{\sum_{i=1}^{10} i \cdot \text{batch blanket surface in } x_i}{\sum_{i=1}^{10} \text{batch blanket surface in } x_i}$$

- X_{trend}: Kind of melting rate of the batch (slope of regression line of the batch coverage in the *x* direction), defined as the change in batch coverage over the *x* direction in meters (Fig. 4).
- Y_{mean}: Average batch blanket coverage over ten bands in the *y* direction in m^2.

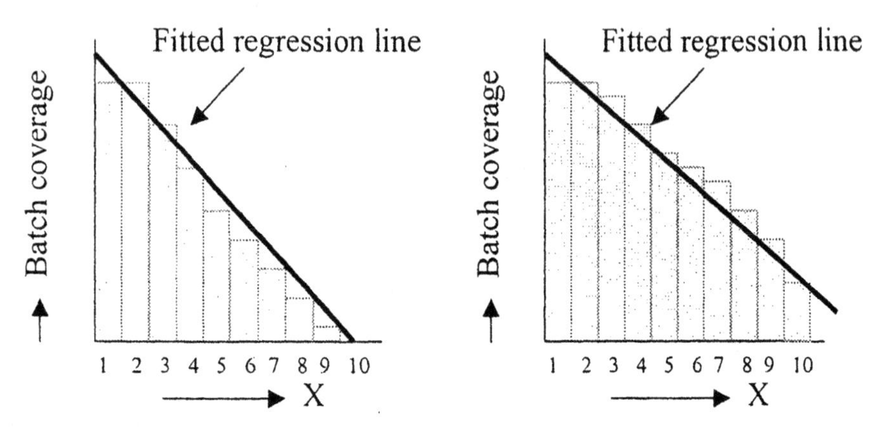

Figure 4. Examples of possible melting rates.

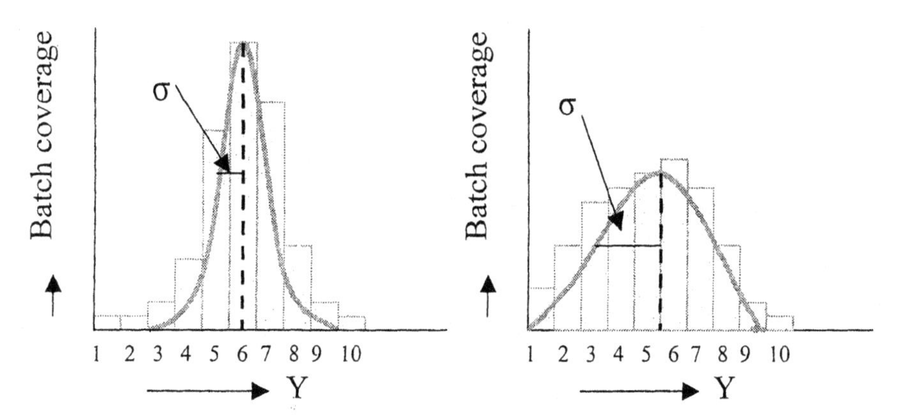

Figure 5. Examples of possible dispersions of batch blanket over the 10 strips in the y direction; σ corresponds with the standard deviation of the fitted curve.

- Y_{center}: Mathematical center of the batch blanket coverage in the y direction (y component of the center of mass) in meters.
- $Y_{standard\ deviation}$: Dispersion of batch blanket in the y direction in meters (Fig. 5).

Results

During tests at a few industrial glass-melting tanks, important correlations between process parameters and some of the previously defined batch blanket parameters have been found. For example, correlations with batch

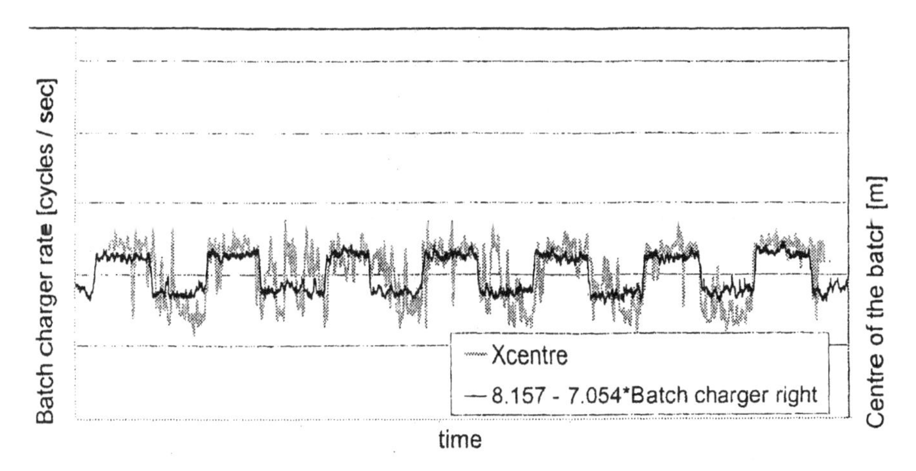

Figure 6. Correlation between the mathematical center position of the batch (gray curve) and the charging rate of the batch charger at the right side of the glass (black curve).

charger rates are found with the symmetry and the determined value of Y_{center}. The charger rates of the two batch chargers (one at the left and one at the right side of the furnace) change over the time; however, the total amount of supplied batch stays constant. The correlation with the batch charger rate and symmetry and Y_{center} are expected results, because the two batch charges are positioned at the sidewalls of the endport-fired tank (see Fig. 3). But less obvious batch blanket parameters such as $Y_{standard\ deviation}$, X_{center}, and X_{trend} also show a good correlation with the feeding rate of one of the batch chargers. For example, the correlation between the mathematical center of the batch blanket in x direction and the batch charger rate is determined as:

$$X_{center} = 8.175 - 7.054 \cdot \text{batch charger rate right}$$

This means that increasing the right batch charger rate results in a decreased batch length (decreased center of mass). However it has to be taken into account that other batch blanket parameters, such as symmetry, will also change by changing the right batch charger rate. Figure 6 shows this correlation. The mathematical center of the batch (X_{center}) is printed as function of time. To show the correlation between X_{center} and the batch charger rate right, the formula shown above is printed as function of time in the same figure and on the same scale.

Besides correlations with batch blanket parameters and batch chargers,

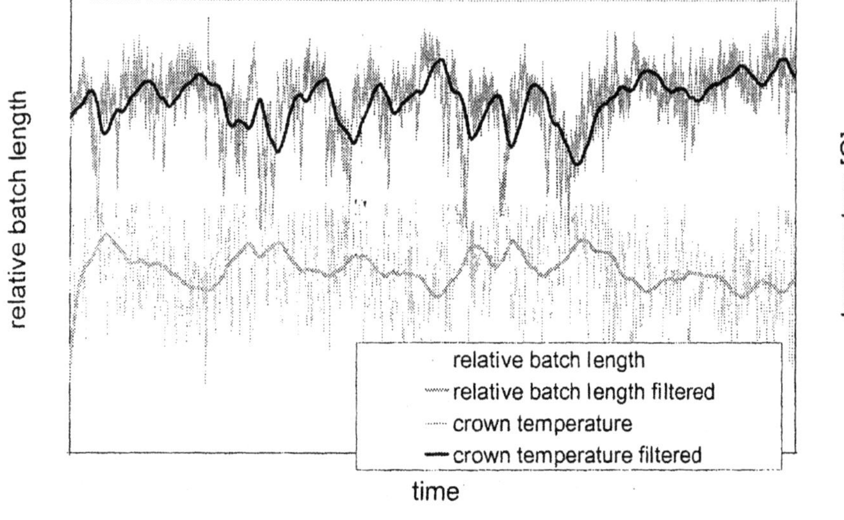

Figure 7. Correlation between (relative) batch blanket length and crown temperature near the bridgewall.

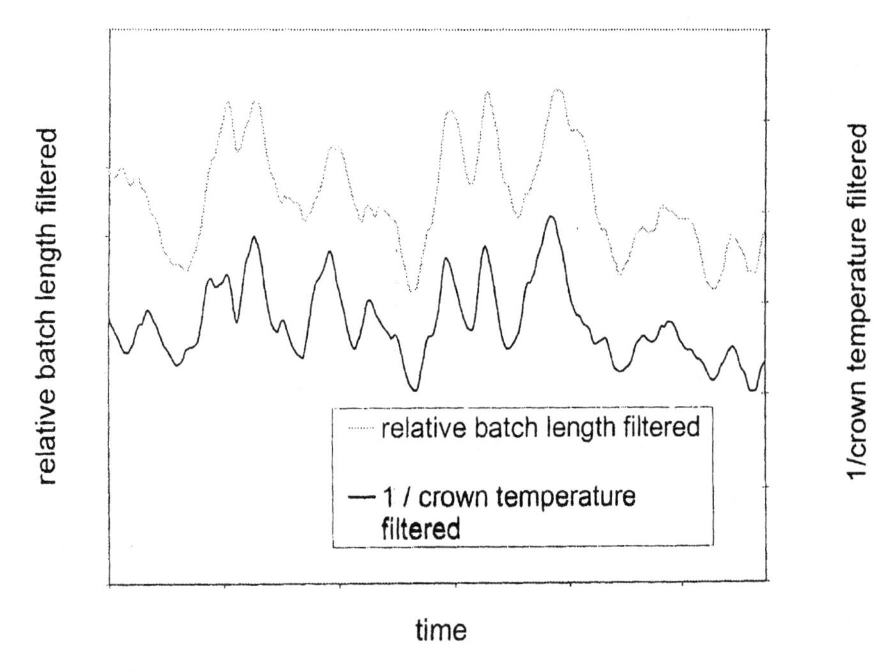

Figure 8. Correlation between (relative) batch length and the 1/crown temperature near the bridgewall.

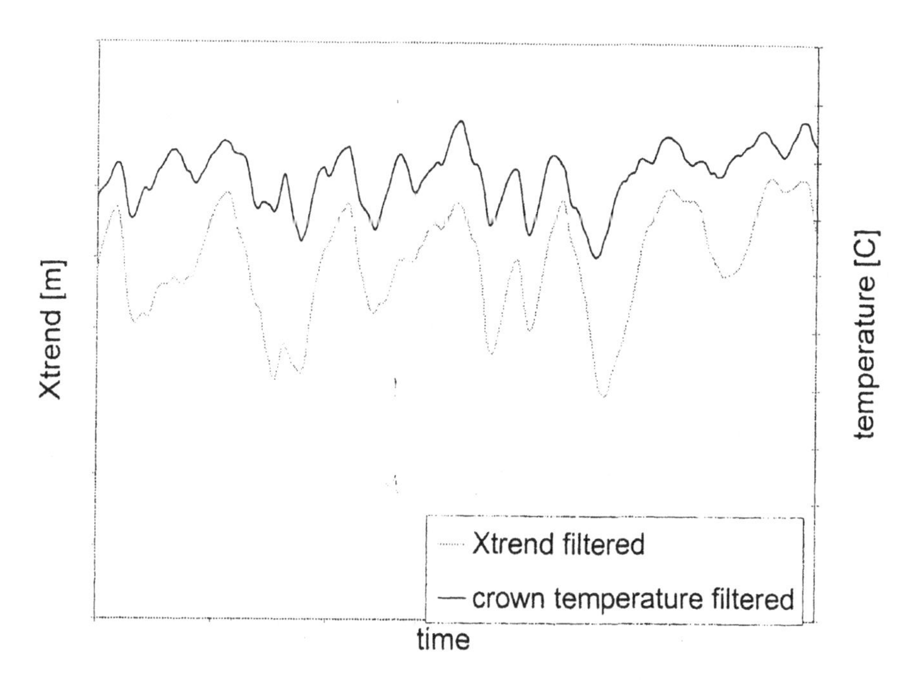

Figure 9. Correlation between the melting rate, X_{trend}, and the crown temperature near the bridgewall.

correlations between batch blanket parameters and crown temperatures are found. Important batch blanket parameters that correlate with crown temperatures are X_{center}, melting rate (X_{trend}), and maximum relative batch length in this specific furnace.

In Fig. 7, the temperature of the crown near the bridgewall and the (relative) length of the batch blanket as function of time are shown. A filter is applied to filter out the high frequencies on the crown temperature and the relative batch blanket length signal. As can be seen in this figure, a longer batch blanket correlates with a lower crown temperature, and vice versa.

To achieve a clear representation on this correlation, the filtered relative batch length and 1/filtered crown temperature are shown as function of time in Fig. 8.

Another example is the correlation between melting rate (X_{trend}) and the crown temperature measured near the bridgewall. The crown temperature near the bridgewall and melting rate are shown as function of time in Fig. 9. An increase in X_{trend} corresponds with an increased melting rate.

An increasing melting rate correlates with an increased crown temperature near the bridgewall.

The determined correlations for this specific furnace can be used to control the batch blanket pattern.

Perspectives

After the determination of clear correlations between selected and measurable batch blanket parameters and process parameters for a specific furnace, the batch monitoring system can be introduced into a furnace control system. Such a coupled system can be used to maintain a stable batch blanket pattern and to control the energy input and the batch feeding rates by the different chargers into the furnace.

Current industrial tests focus on finding correlations between batch blanket parameters and glass quality characteristics, such as blisters and seed counts.

Thermal Imaging of All Furnace Internal Surfaces for Monitoring and Control

Serguei Zelepouga and David Rue
Gas Technology Institute

Ishwar Puri
University of Illinois

Ping-Rey Jang and John Plodenic
Diagnostic Instrumentation and Analysis Laboratory, Mississippi State University

John Connors
PPG Industries, Inc.

Introduction

Optimized glass furnace operation and control depend on accurate temperature data for critical glass and refractory points and areas in the melter combustion space. Infrared thermal imaging sensors provide significant advantages over traditional point thermocouples. While thermocouples are reliable, they can provide only limited temperature information. Thermal imaging sensors can provide the same information at critical points while also providing a temperature map of the complete combustion space. This real-time information can be used to monitor melter temperature profiles, refractory hot spots, location and movement of the batch line, and so on, and to provide more complete temperature data to the melter controller. Several commercial thermal imaging systems are available for glass melter use, and two research groups, the Diagnostic Instrumentation and Analysis Laboratory (DIAL) at Mississippi State University and the Gas Technology Institute, are developing low-cost thermal imaging systems to provide simultaneous real-time thermal maps of all glass and refractory surfaces along with real-time visual video for combustion monitoring.

Temperature Measurement

In many locations in a glass factory, precise temperature data are needed for process control. Other measurements are needed for process monitoring, trending, and safety. The wide range of temperatures and different environments necessitates careful selection of appropriate thermocouples and other temperature measuring devices. High temperatures and harsh environments make critical melter temperature measurement particularly

Table I. Temperature sensors

	Thermocouples	Pyrometers	Thermal imaging
Available?	Yes	Yes	Limited
Cost	Low	Moderate	High
Accuracy	Good	Good	Good
Use	Easy	Easy	Moderate
Positioning	Extensive	Viewport	Viewport
Contact	Needed	None	None
Output data	Point	Point	Field
Data quantity	Low	Low	High
Standardized?	Yes	Partly	No
Controls compatible	Yes	With effort	With effort

challenging. Several types of melter temperature sensors are available, including type R and other platinum thermocouples, handheld pyrometers, infrared thermal imaging systems, and specialty devices such as the INEEL-AccuTru temperature monitor.

A comparison of melter temperature sensors is shown in Table I. Thermocouples are available, well understood, reliable, and well suited for supplying data to control systems. Pyrometers are also available and provide noncontact point temperature readings at a somewhat higher cost. Infrared thermal imaging systems have only limited market presence today and cost more, but these sensors offer the capability to measure a field temperature map. The large amount of data from thermal imaging systems must be processed by appropriate software to obtain needed information. This software is a key part of any thermal imaging system used in industry.

Thermocouples have accuracies only to within several degrees, and accuracy can be hurt by several causes, including iron impurity[1] and reactions with insulation materials. A number of platinum thermocouples have been found to drift with prolonged use.[2] Thermal imaging systems avoid these problems by measuring the intensity of infrared radiation.[3] Black body radiation is a maximum in the infrared part of the spectrum. The peak black body intensity rises and occurs at shorter wavelengths as surface temperature increases. There is no universal thermal imaging system. Instead, the detector is chosen based on temperature, desired wavelength(s) for measurement, and avoidance of spectral regions in which radiating species might interfere with readings.

Thermal imaging systems and pyrometers rely on Planck's Law to relate monochromatic light intensity $I_{i,b}$ at a known wavelength λ_i to the absolute temperature T of the radiating surface.

$$I_{i,b}(\lambda_i, T) = c_1/\{\lambda_i^5 [\exp(c_2 / \lambda_i T) - 1]\}$$

where c_1 and c_2 are the first and second radiation constants.

In practice, filters or gratings are used to generate the monochromatic light and Planck's Law can be rewritten as follows:

$$I_{\lambda i}(x,y) = c_{\lambda,i}\, d_{\lambda,i}\, \varepsilon_{\lambda,i}\, I_{\lambda,i}(\lambda i, T)$$

where $c_{\lambda,i}$ is the spectral response at λ_i; $d_{\lambda,i}$ is the filter transmission; and $\varepsilon_{\lambda,i}$ denotes surface emissivity at λ_{1i}.

In single wavelength infrared thermometry, the surface emissivity must be known to relate temperature to observed intensity. Surface emissivitiy data may not be available, and different surfaces in a map will have varying emissivities. The need to know surface emissivity data can be eliminated by acquiring intensity data at two close narrow wavelengths.

$$\lambda_{1,1}(\mathrm{x,y}) = c_{1,1}\, d_{1,1}\, \varepsilon_{1,1}\, c_1 / \{\lambda_1^5 [\exp(c_2 / \lambda_1 T) - 1]\}$$

$$\lambda_{1,2}(\mathrm{x,y}) = c_{1,2}\, d_{1,2}\, \varepsilon_{1,2}\, c_1 / \{\lambda_2^5 [\exp(c_2 / \lambda_2 T) - 1]\}$$

When λ_1 and λ_2 are close, a reasonable assumption can be made that $\varepsilon_{\lambda 1}$ and $\varepsilon_{\lambda 2}$ are the same. Solving these two equations simultaneously produces a nonlinear equation independent of emissivity. Iterative methods are often used to solve the nonlinear equations relating temperature to intensity. Software is a key component of any thermal imaging system. The choice of wavelengths (from 0.6 to 4 μm), the iterative solution approach, the speed and accuracy of the solution method, and the format of the temperature map output are parameters that must be considered when selecting and using a thermal imaging system.

Commercial Thermal Imaging

In practice, a thermal imaging system used on a glass-melting furnace must include several components. These include, at a minimum, a lens system, a detector with or without a shutter, a frame grabber to capture and save images, and software to produce temperature maps from intensity maps.

Software is specific to each manufacturer and is often tailored to the needs of specific applications. For that reason, the details of available software are constantly evolving and should be evaluated by reviewing the offerings of thermal imaging system suppliers. Because thermal imaging systems capture infrared light, the lenses used must transmit light at the appropriate wavelengths. Germanium and sapphire lenses with special coatings are commonly used for infrared thermometry. Less expensive silicon lenses can be used at wavelengths above 1 μm but are not transparent at lower wavelengths. High-temperature measurements (such as glass melters) can be conducted in the 0.6–1 μm range because black body radiation intensity is near the maximum in this region at high temperatures. Typical optical glasses can be used for lenses in the 0.6–1 μm region.

Two types of detectors are used. Linescanners use a one-dimensional array while focal plane arrays (FPAs) use a two-dimensional detector array. Linescanners are less expensive than FPAs, but a scanning mechanism is required if a two-dimensional temperature map is desired. Linescanners are ideal for creating the thermal image of a moving object or for scanning in a line across a surface (such as across a molten glass surface to detect the batch line). Focal plane arrays acquire all data simultaneously with resolutions of 120 × 120 to 1000 × 1000 or more. A list of some of the currently available linescanner and FRA thermal imaging systems, manufacturers, and ranges of use is shown in Table II.

Sensors used on linescanners and FPAs are either microbolometers or semiconductor photon detectors. Microbolometers are devices that measure the temperature increase resulting from photon absorption. These devices can be quite accurate and external cooling is not a requirement. Array resolution, however, may be low. Microbolometers are most often used to measure surface temperatures below 500°C, but they can be used at higher temperature with the addition of neutral density filters. Semiconductors measure the voltage that results from the release or transfer of electrons caused by photon absorption. FPAs are also accurate. Thermoelectric or other cooling is used to reach even higher levels of accuracy. Much higher resolutions can be achieved with these detectors. Semiconductor materials are selected based on temperature ranges of interest, accuracy required, and cost. Common materials include INGaAs, HgCdTe, InSb, and silicon. Silicon is transparent at wavelengths above 1 μm and so must be used only for near-infrared measurements.

Table II. Available thermal imaging systems for all applications

Detector type	Manufacturer	Temperature range (°C)	Spectral range (μm)
Thermal imagers			
Uncooled microbolometer FPA	Land, Flir, Mikron, etc.	(20)–500	7–14
Thermoelectrically cooled HgCdTe FPA	Land, Flir, Mikron, etc.	(20)–2000	3.2–4.2
Thermoelectrically cooled InGaAs FPA	Indigo, etc.	250–1000	0.9–1.7
Thermoelectrically cooled InSb FPA	Indigo, etc.	0–350	1–5.4
Linescanner			
Thermoelectrically cooled IR detector	Land, Flir, Mikron, etc.	0–1400	1.1–14
Thermometer			
Infrared	Land, Flir, Mikron, etc.	(40)–3000	0.8–1.8

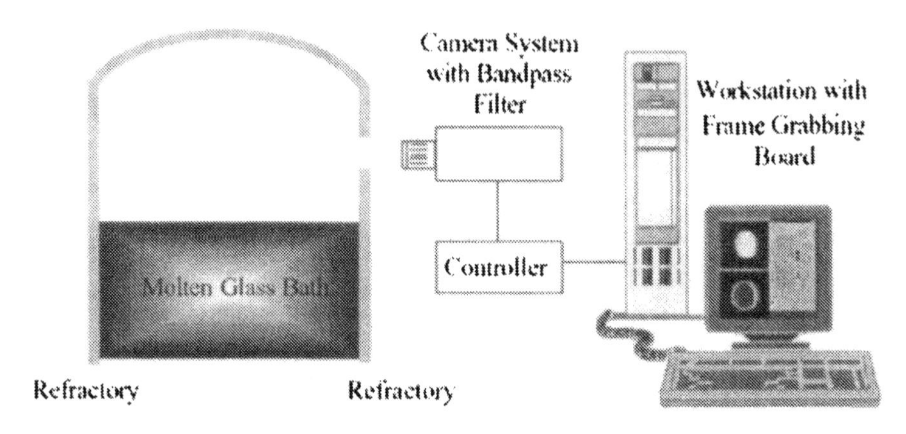

Figure 1. Simplified mechanics of glass melter thermal imaging.

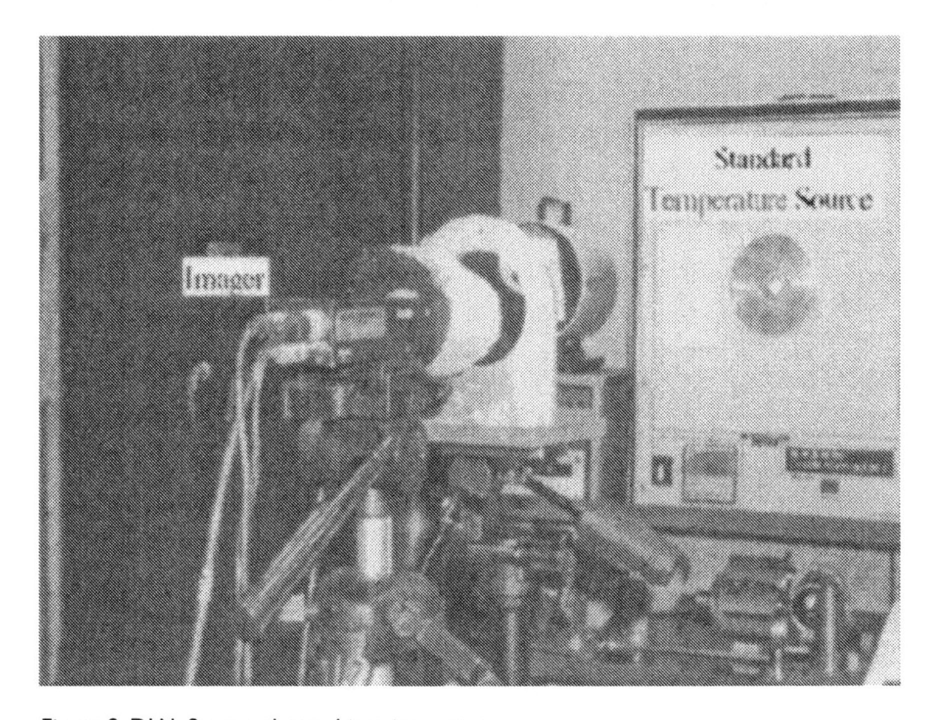

Figure 2. DIAL 2-wave thermal imaging system.

Glass Melter Thermal Imaging

Both DIAL and GTI have demonstrated prototype thermal imaging systems on working glass melters. An overview of the experimental approach taken by both groups is shown in Fig. 1. A water-cooled periscope with proper

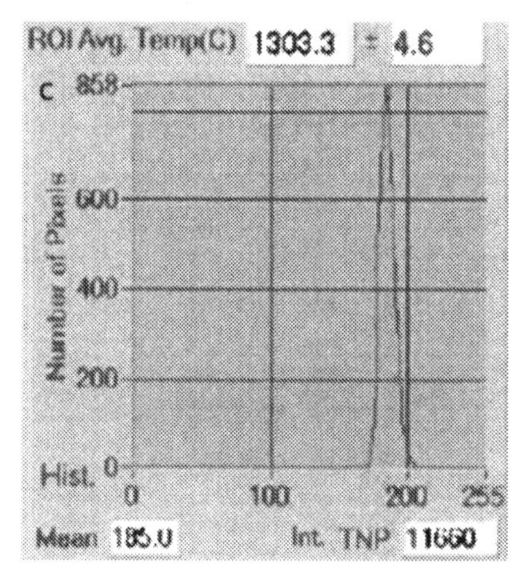

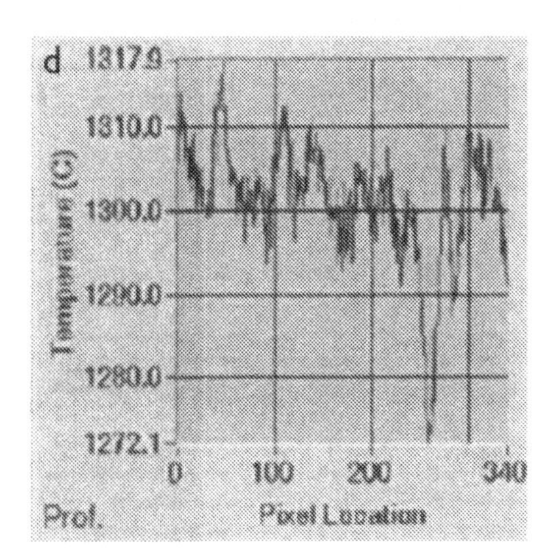

Figure 3. Thermal imaging of Techniglas melter: (a) refractory monochrome image, (b) false color thermal map of refractory, (c) histogram, (d) line profile.

optics was inserted through various melter viewports. Two narrow bandpass filters produced monochromatic near-IR light measured for intensity by silicon focal plane arrays. The intensity maps were saved to a computer by a frame grabber card, and proprietary software was used to convert intensity maps at the two close, narrow wavelength bands into a thermal map.

Figure 2 shows a photograph of the DIAL 2-wave thermal imaging system camera in position to be calibrated using a standard black body source. Components included a near-IR digital camera, a filter wheel for two filters between 0.5 and 1 μm, a water-cooled periscope, a motorized zoom lens, and a dedicated computer with frame grabber card. Software was designed for image acquisition and processing, image false color selection, temperature-color bin assignment, image playback simulation, and interactive region and line profile selection. Figure 3 shows thermal imaging examples

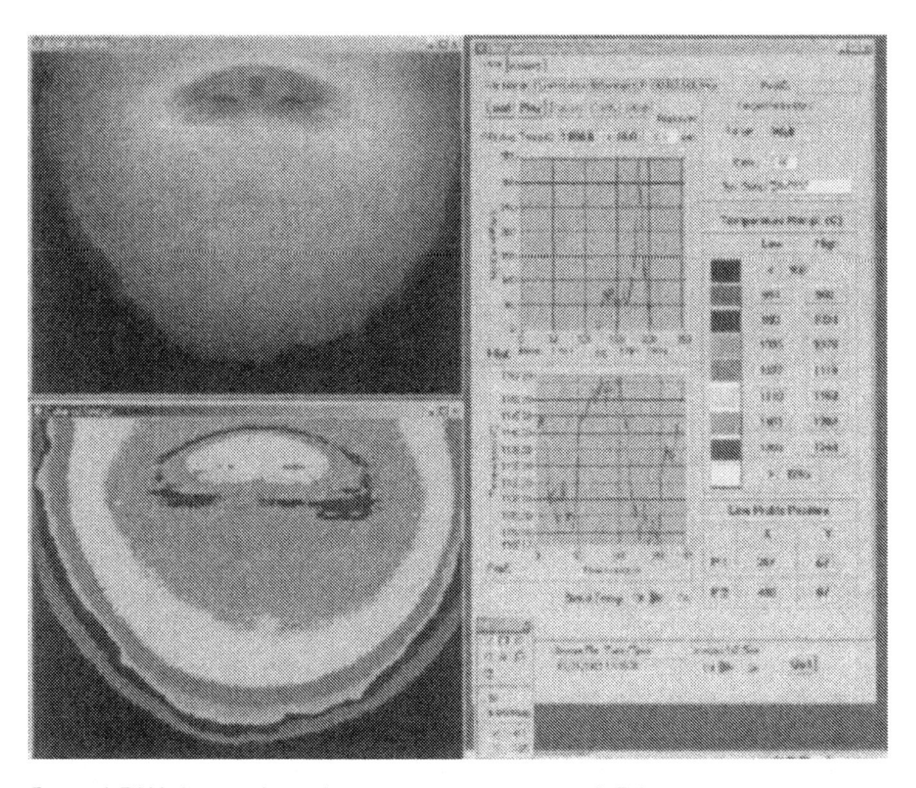

Figure 4. DIAL 1-wave thermal imaging system output with Eclipse camera input.

from the DIAL 2-wave thermal imaging system acquired during testing on a Techniglass furnace. Wall hot and cold spots are clearly visible visually and on the histogram and the line profile.

DIAL researchers also developed a 1-wave thermal imaging system compatible with Eclipse melter camera systems. The 1-wave system required calibration against a pyrometer. Output from this system looking toward the charging end of a melter is presented in Fig. 4. Positions of the batch and bubble line are clearly visible, and refractory and glass surface temperatures are easily seen in the thermal image. Flame positions are also distinct.

A photograph of the GTI 2-wave thermal imaging system taken during data collection at a PPG Industries oxy-gas float furnace in Meadville, Pennsylvania, is presented in Fig. 5. GTI used a different near-IR camera and a filter wheel with two filters (0.50 and 0.55 μm in this case). The device, like the DIAL instrument, employed a water-cooled periscope,

Figure 5. GTI 2-wave thermal imaging system.

appropriate lenses, and a frame grabber card in a computer. GTI, like DIAL, developed software for image capture, storage, and analysis. The GTI software allows temperature averaging and temperature determination for any defined point, line, or area.

Figure 6 shows images taken by GTI engineers during testing at PPG. One image is a monochrome intensity map providing the field of view along with the position of a flame, the walls, the glass surface, and unmelted batch material. The two thermal image maps are all processed from the

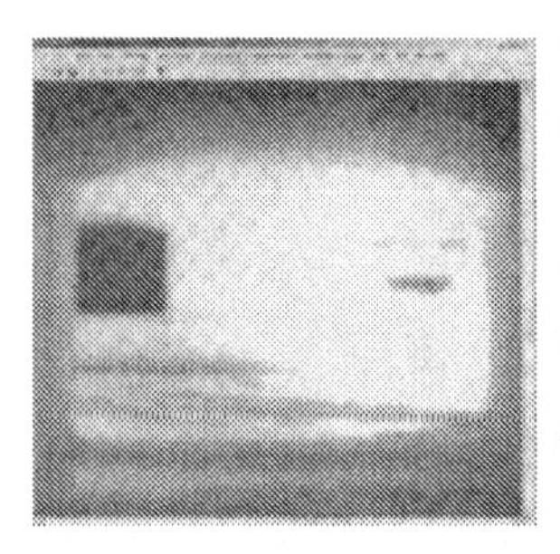
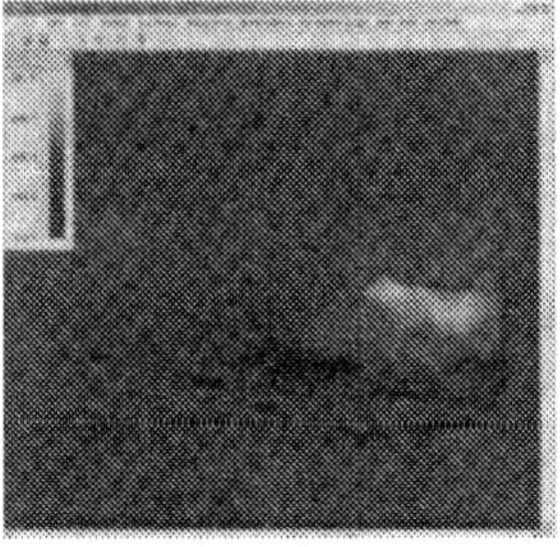
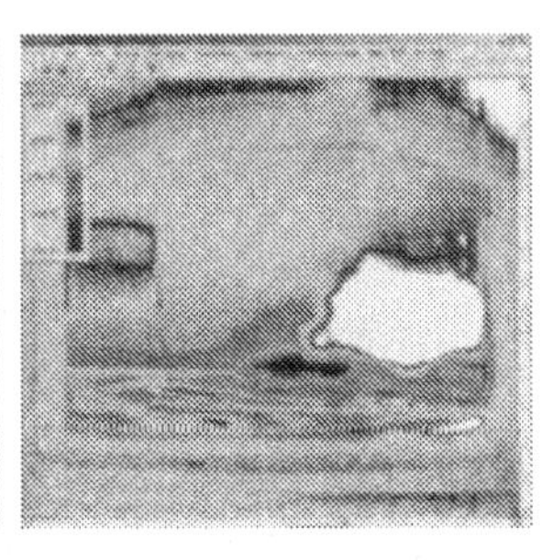

Figure 6. GTI thermal maps of PPG melter: (left) intensity map at 0.55 μm, (center) thermal map in the 925–3315°C range, (right) thermal map in the 925–1650°C range.

same two intensity maps. Different temperature scales are set for each image. The true color thermal map covering 925–3315°C shows the position of the flame, the glass, and the walls, but no detailed temperature data can be collected. The false color thermal map covering 925–1650°C provides a good deal of temperature information. The flame blocks part of the map, but the walls, glass surface, and batch material are all present. All of these data are available to the operator and can be used in any number of ways to assess melter operation and to compare operation with intended or expected behavior.

Thermal Imaging Beyond the Melter

Thermal imaging systems offer opportunities for improved data acquisition and process control in other parts of the glass-making operation. The technique can be used for glass defect determination and melter inspection in all glass industry segments. Locations of particular interest for thermal imaging in the container glass plant include the forehearth, gob formation, and container forming. In the flat glass plant, thermal imaging can benefit operation of the tin bath, the annealing lehr, and the processes for tempering, shaping, and bending. Linescanner thermal imaging equipment has been shown to easily identify defects in flat glass ribbons during passage along the tin bath. Fiber plant areas of interest for thermal imaging use include the forehearth and the bushings. Specialty glassmakers will benefit from thermal imaging systems in gob formation and in product forming.

Thermal imaging systems can be valuable tools in monitoring glass melters, supporting melter design calculations, and validating modeling predictions. An issue that must be addressed in working with these systems is that large amounts of data must be acquired and processed on a regular basis. Data acquisition and control systems do not yet have standardized approaches for handling and using these data. Flexibility must be built into software to use temperatures of batch line position, hot and cold sports, and critical refractory points as information for process monitoring and control. A final consideration is the cost of thermal imaging systems. Systems today cost between $40 000 and $100 000, making them too costly for dedication to individual glass melters. Costs, however, are declining, and GTI estimates the commercial version of their system will market in a few years for under $15 000. This is primarily because the necessary hardware is declining in cost and becoming more readily available. Also, as markets grow for thermal imaging systems, costs are expected to decrease with increased sales volume.

References

1. B. E. Walker, C. T. Ewing, and R. R. Miller, "Study of Instability of Noble Metal Thermocouples in a Vacuum," *Rev. of Sci. Instruments,* **36** [5] 601–606 (1965).
2. A. D. Watkins and A. Thome, "High Temperature Thermocouple Degradation Study." G-Plus Final Report, 2003.
3. G. M. Carlomango and L. de Luca, "Infrared Thermometry for Flow Visualization and Heat Transfer Measurements"; presented at the International Conference on Engineering Education, Rio de Janiero, Brazil, 17–20 August 1998.

Improvement in Glass Blister Quality by Throat Design

R. R. Thomas
Corning Incorporated, Corning, New York

The throat of a glass furnace is sometimes the life-limiting part of the melting process. For Corning's color TV glass panel tanks, improvements in throat life achieved by using a submerged throat design were offset by poorer blister quality. This paper discusses the design of a standard throat that produced improvement in blister quality using return flow and at the same time improved throat life.

Introduction

Improving glass quality and increasing glass furnace life are two powerful ways to reduce the cost of making glass. Frequently, however, they have an inverse relationship. For example, improving glass quality by eliminating gaseous inclusions through increased glass bath temperature usually causes the tank life to be decreased. Generally, it is rare to be able to create a design that will improve the melting process in both ways. However, furnace engineers at Corning Incorporated recognized the crucial role that the throat between the melting zone and the refining zone plays in both furnace life and glass quality. By making relatively small design changes, they were able to bring about improvements in both areas. This paper discusses learning about the role of throats in creating good blister quality accumulated over three campaigns of one of Corning's CTV panel tanks at its State College, Pennsylvania, plant.

Procedure

The demand for ever-larger color TVs has forced cathode ray tube glass manufacturers to continuously improve the inclusion quality of their TV panel glass. Gaseous inclusions (blisters) have historically been the largest class of rejectable inclusions for Corning's TV panel manufacturing process. Ten years ago, acceptable manufacturing efficiency could be obtained for 27-in. color TV panels (weighing up to 15.4 kg) with blister counts less than about 0.033 per kilogram. Presently, for 38-in. panels weighing in excess of 45 kg, blister levels must be less than 0.011 per kilogram for reasonable select levels.

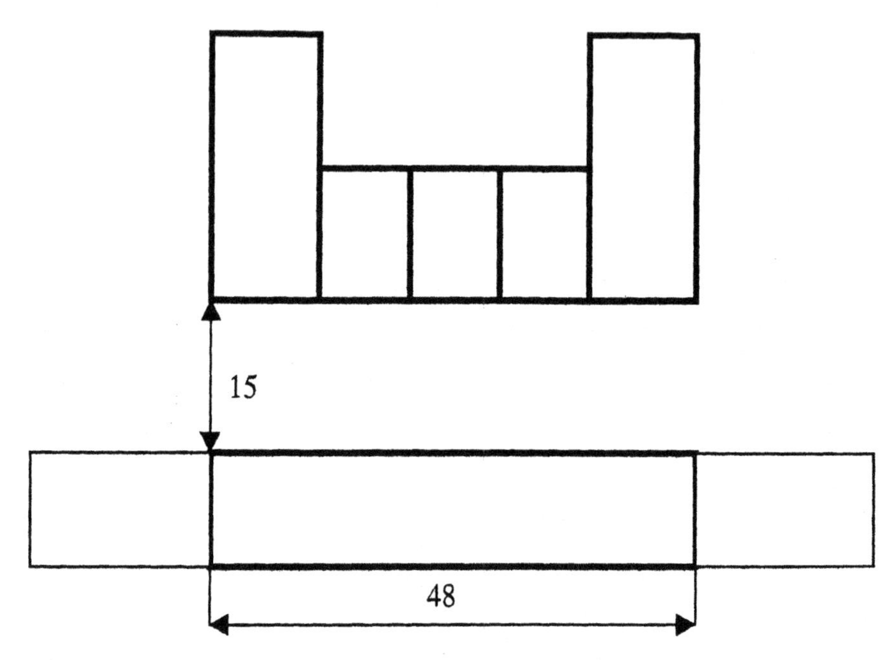

Figure 1. Schematic view of standard throat (width = 24 in.).

Concurrently, pressure for cost reduction has driven the need for increased melting furnace life. Corning's experience with TV panel furnace life has been that the throat between the melt zone and the refining zone is frequently the life-limiting part of the process. Corning has used two throat designs over the last 20 years: the standard throat and the submerged throat. The main difference between them is that the bottom of the standard throat is even with the melt zone and refining zone bottoms whereas the bottom of the submerged throat is somewhat below the level of the other bottoms. In both cases, the glass line is above the top of the throat (bottom of the throat covers).

Figures 1 and 2 show schematic views of a standard and a submerged throat, respectively. Each design has facer blocks at each end that form part of the front wall on the melt side of the throat and the back wall of the refiner on the refiner side of the throat. In between the melt facer and the refiner facer, the throat consists of side blocks (stringers) and top blocks (covers). Generally, the top facer blocks and the covers are the most heavily worn parts of the throat and historically have been the areas that caused a melting campaign to end.

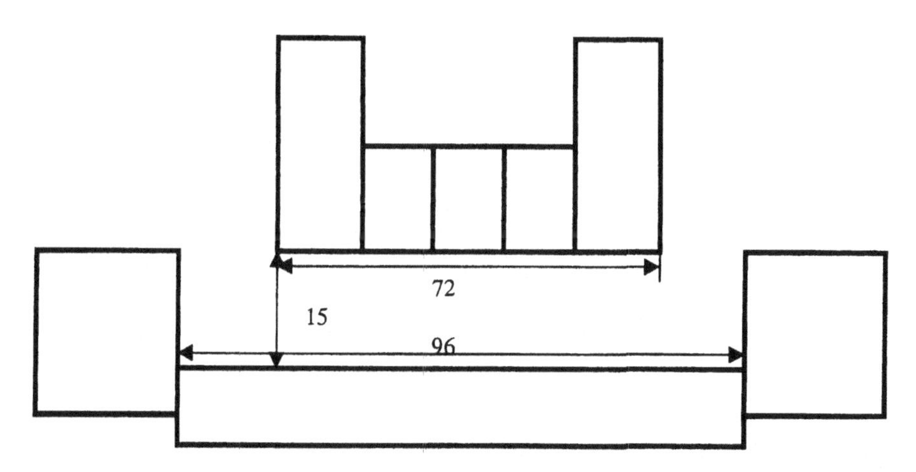

Figure 2. Schematic view of submerged throat (width = 24 in.).

Figure 3. Assembly view of standard throat before installation.

Figures 3, 4, and 5 show a standard throat in three views: Figure 3 is the assembly before installation, Fig. 4 is the new throat installed, and Fig. 5 shows the same throat worn at the end of a five-year campaign. The assembly view shows the melt side facer, covers, and stringers. The bottom of the

Figure 4. Installed view of new standard throat (Campaign 2).

Figure 5. View of standard throat at end of Campaign 2.

Table I. Throat dimensions (mm)

	Standard throat		Submerged throat	
	New	End of campaign	New	End of campaign
Height	381	635	381	483
Width	610	686	610	622
Length	1219	1219	1829	1829

throat is at the same level as the bottom of the furnace. The throat dimensions are shown in Table I.

The throat shown in this table had a life of five years. The wear rate of the covers was approximately 38 mm per year. This high wear rate prompted an investigation into ways that the life of the throat could be improved. Experience with other melters at Corning suggested that a submerged throat would produce a significant improvement in life.

The TV panel tank at Corning's State College plant was rebuilt with the submerged throat shown in Fig. 2. The submerged throat delivered the desired life extension. The life improved from 5 to 6.5 years. The throat actually could have lasted longer than 6.5 years. The tank was repaired for other reasons. Figure 6 shows a view of the throat from the refiner side at the end of the campaign. The large notch out of the top of the throat was a spall from the facer. The throat covers were worn less than 4 in.

However, for the campaign with the submerged throat (Campaign 1), glass blister quality was poor and subject to upsets during the campaign with the submerged throat. Figure 7 shows blister quality during the campaign, averaging over 0.044 per kg (.020 per pound) for most of the campaign and jumped to more than 0.075 near the end of the campaign. The ratio of good panels to total possible panels was less than desired during the campaign with a submerged throat and it was decided to replace the submerged throat with a standard throat.

A standard throat with the same dimensions shown in the Table I was installed for Campaign 2. Figure 8 shows the blister level throughout Campaign 2 compared with Campaign 1. Two observations were made from the comparison of Campaigns 1 and 2. First, glass blister quality was generally better in Campaign 2 than in Campaign 1; and second, glass blister quality generally got better as Campaign 2 progressed. These observations prompted further investigation.

Figure 6. View of a submerged throat at end of Campaign 1.

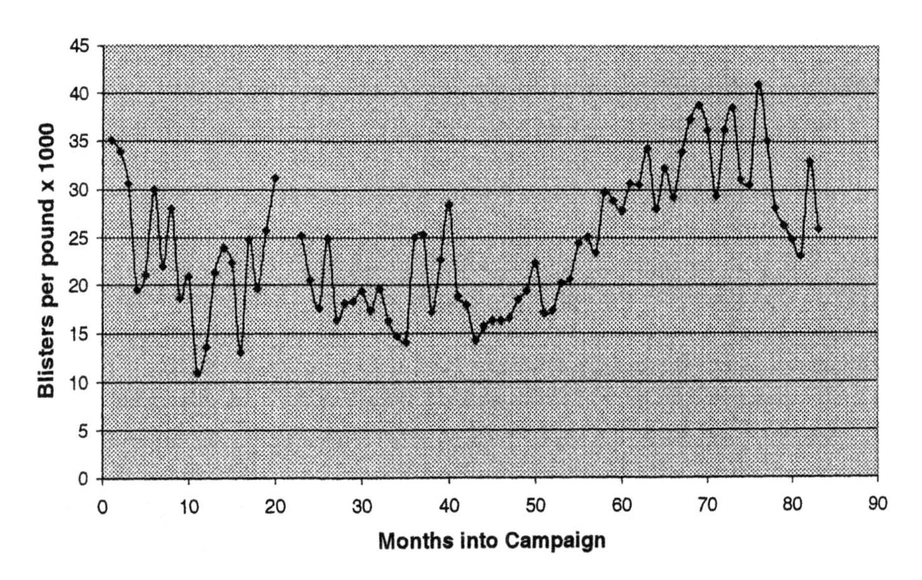

Figure 7. Blister quality for Campaign 1.

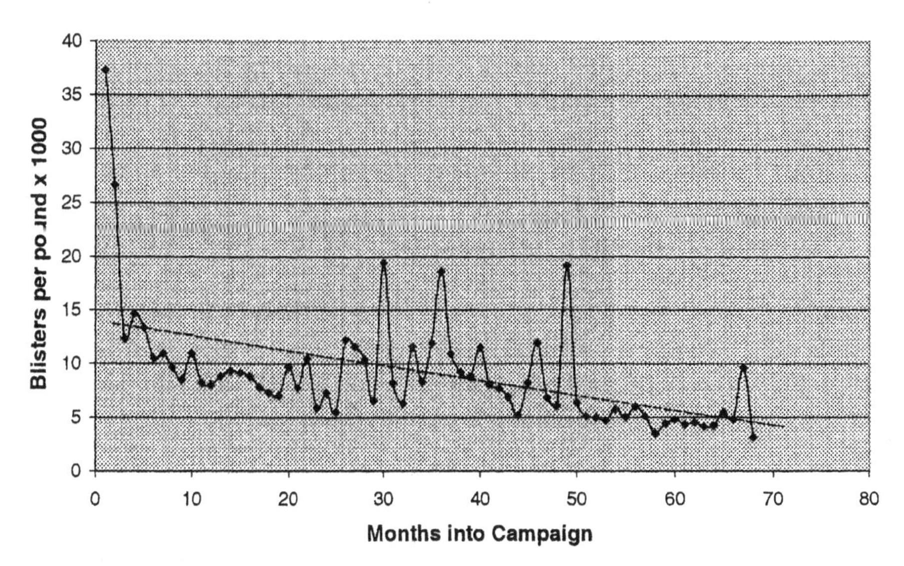

Figure 8. Blister quality for Campaign 2.

Discussion

It was clear in comparing Campaigns 1 and 2 that the gain in blister quality with the use of a standard throat in Campaign 2 was somewhat offset by a reduction in life. What was not clear was what mechanism caused the improvement in blisters.

Data from a thermocouple placed on the melter bottom near the inlet to the throat provided the clue to the relationship between the standard throat and improved blister quality. At the beginning of Campaign 2, the glass entered the throat from the melt end at a temperature of about 1430°C (thermocouple CBM), exiting into the refiner at a temperature of about 1280°C (RETH). The refiner bottom temperature (RB) was about 1140°C. Almost immediately after the start of the campaign the CBM temperature began to decline. The CBM temperature steadily declined throughout the campaign to a temperature of about 1300°C by the end of the campaign while RETH and RB were held constant.

This behavior can be explained by the concept of return flow. Stanek[1] explained the phenomenon of return flow in a throat in 1977. He showed that if the temperature at the inlet (melter) end of the throat was sufficiently greater than the temperature at the outlet (refiner) end of the throat, it is possible to set up a return current in the throat. The forward flow of glass that is due to the tank pull would travel through the upper part of the throat and a portion of that glass will sink to the bottom of the refiner and be drawn back through the bottom part of the throat into the melter. Furthermore, he showed that under the right conditions, the return flow can actually exceed the pull. The forward flow in the upper part of the throat would be the pull plus the return flow.

The steadily decreasing CBM thermocouple is explained by the gradually increasing return current of relatively cold glass passing through the bottom half of the throat and cooling the CBM as it mixes with hot glass from the melter. The improvement in blister quality then would be caused by returning increasing amounts of glass from the refiner back into the melter for more residence time at fining temperatures. It is almost impossible to get return flow with a submerged throat. The CBM thermocouple for the submerged throat in Campaign 1 did not change appreciably during the campaign.

This trend of steady improvement in blister quality during a campaign was confirmed on another panel tank with a standard throat, prompting the desire to somehow be able to build a new tank with the properties of an old tank. It now became possible to do that by designing the throat to have return flow on day one of the campaign.

One of the objectives of the melter design for Campaign 3 was to produce blister quality of 0.022 per kilogram (or better), the quality level produced by the tank after about one year of Campaign 2. Another was to increase the fill rate from 172 335 to 204 080 kg/day. The throat had to be designed without the benefit of inspection of the worn throat from Campaign 2 in order to have time to prepare drawings and procure materials. So the first task was estimating the size of the throat and the amount of return flow achieved roughly one year into Campaign 2.

From measurements taken at cold inspection of other standard throats, an estimate was made of the throat dimensions 1 year into the campaign. It was estimated that the throat dimensions at that time were 622 mm (24.5 in.) wide and 432 mm (17 in.) high. These dimensions along with glass properties and operating temperatures were used with the Stanek method to estimate the return flow. This calculation is shown in Fig. 9. The return flow was estimated to be about 7.5% of the pull. This became the target return flow for the 204 000 kg/day tank.

For the enlarged tank for Campaign 3, the same calculations were done with the objective of achieving about 7.5% return flow on day one. These

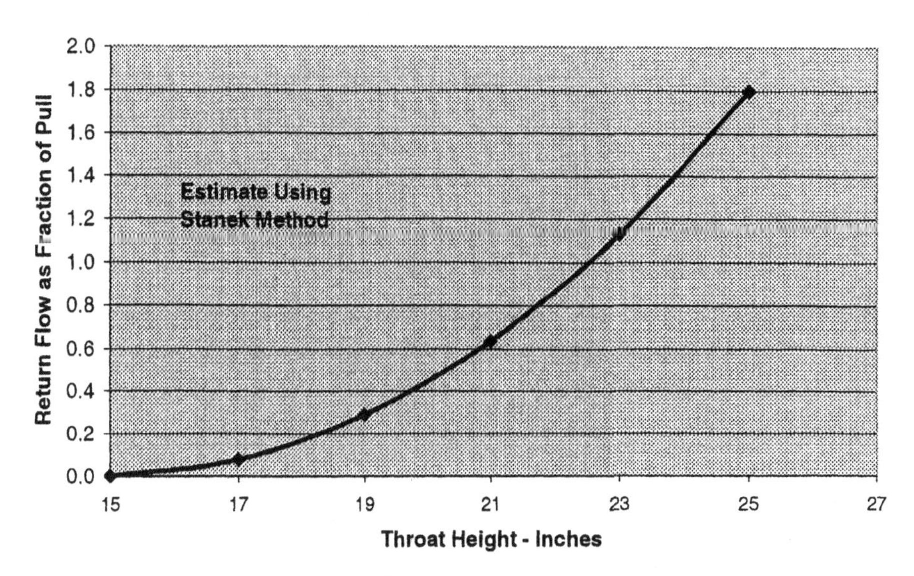

Figure 9. Return flow for Campaign 2.

calculations are shown in Fig. 10. Some considerations in the choice of throat dimensions to achieve the 7.5% return flow on day one were:

1. Return flow is proportional to roughly the third power of throat height and the first power of throat width for the same temperature difference.
2. Throat wear is at least three times as fast for the covers (height) as for the sides (width).
3. Cover wear will determine the life of the throat.
4. A wider throat will create longer spans for the covers. Because this was a repair of an existing tank, a wider throat also created changes in glass contact refractory and steel for the entire front wall of the tank.

From the data in Fig. 10, it was determined that several throat configurations would meet the requirement for day one return flow. Based primarily on point 4 above, the configuration chosen was 432 mm (17 in.) high × 610 mm (24 in.) wide × 1219 mm (48 in.) long. This combination predicts a return flow of about 6% on day one. However, a wider configuration provides more thickness for the covers, so a configuration of 406 mm (16 in.) high × 762 mm (30 in.) wide × 1219 mm (48 in.) long could add almost a

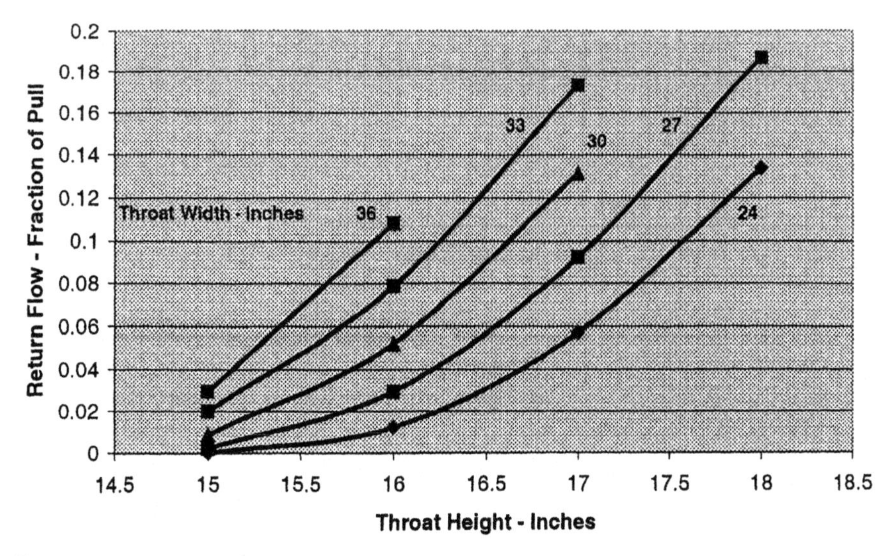

Figure 10. Return flow for Campaign 3.

year to the throat life. A throat wider than 30 in. for this application would add significant complexity with very little incremental life. Had the consideration of changes in the rest of the front wall not been important, the decision would have been for a 406 mm (16 in.) high × 762 mm (30 in.) wide × 1219 mm (48 in.) long throat.

The resulting blister quality for Campaign 3 is shown in Fig. 11. Blister counts at the start of the campaign were about 0.13 per kilogram and stayed between 0.04 and 0.13 per kilogram for the rest of the campaign. The CBM thermocouple confirmed that return flow was taking place because it was much colder than upstream bottom thermocouples.

Conclusion

Return flow brings glass from the refiner back into the high-temperature fining zone in the melter so that blisters from all sources (melting, refractory outgassing, etc.) have another opportunity to fine out of the glass. The concept of return flow contributed significantly to the effort to reduce gaseous inclusions in TV panel glass. Understanding the role of throat height and width can enable a furnace design team to take advantage of return flow and at the same time use width to minimize throat height, where throat wear is highest.

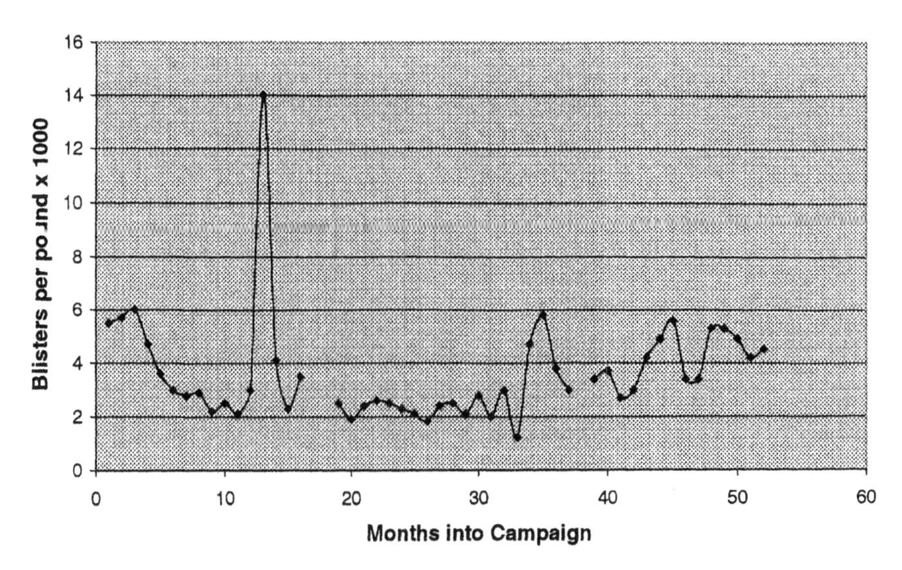

Figure 11. Blister quality for Campaign 3.

Reference

1. Jaroslav Stanek, *Electric Melting of Glass.* Elsevier Scientific Publishing Company, 1977. Pp 177–192.

Emerging Areas

Overview of the Activities of the Technical Committees of the International Commission on Glass

Henk de Waal
International Commission on Glass, Delft, The Netherlands

The International Commission on Glass (ICG), founded in 1933, in 2003 had member organizations in 32 countries. Every three years an International Congress on Glass is held; in other years an annual meeting is organized by one of the member organizations. The technical committees form another highly important part of ICG's activities. Supervised by the Coordinating Technical Committee, there are 21 active technical committees, totaling more than 350 members who are experts in their particular fields drawn selectively from universities, institutions, and industry. With more than 50% of the technical committee members employed in the glass industry, this structure provides an excellent vehicle for communication and exchange of information between industry and academia. The activities of a technical committee may cover a wide range, depending on the demand in a certain area of investigation, and can certainly vary with time. The main activities, however, are discussions on selected topics among the members; round robin tests to establish (the accuracy and reliability of) measurement, analytical, or modeling methods; development of new standards; literature surveys; organization of topical sessions at international conferences; and short training courses and workshops, mostly on technological subjects. An overview is presented of the areas of interest of all committees, which vary from basic science to purely technological subjects. Also, details of recent activities of each technical committee are discussed.

Introduction

The International Commission on Glass, founded in 1933, has from the very beginning had a keen interest in matters concerning exchange of information and education and training in the field of glass science and technology. One of the ways ICG accomplishes this objective is the organization of an International Congress on Glass every three years in one of the member countries; in other years an annual meeting is organized by one of the member organizations.

In its first official document, the declaration signed by the six founding members on 23 September 1933, one of the objectives of the ICG is stated as "to be a clearing-house of international, technical and scientific work."

When relations were reestablished after World War II, a new constitution was approved by the 13 member organizations in 1950, stating that it was the object of the ICG "to promote and stimulate understanding and cooperation between different countries for the exchange of information on the art, science and technology of glass by such means as serving as an international center for the exchange of information and assisting everywhere in the development of interest in glass."

Still, it was only at the eighth annual meeting of the ICG, in July 1957, that the organization of collaboration in R&D at international level took its final form. It was proposed to establish a special committee on science and technology with five subcommittees to organize joint activities in the following fields:

- Definitions and terminology.
- Chemical durability.
- Glass quality.
- Temperature-viscosity relations.
- Furnace operation (with particular reference to temperature measuring methods and targets of measurement).

The aims were essentially to improve relations between the ICG and the glass community, to extend the coverage of glass science, and to set aside time during congresses to communicate the results of the work undertaken. The different subcommittees were encouraged to organize the publication of their results in the form of extensive literature reviews or monographs.

No less than 38 publications thus appeared between 1957 and 2003.[1] This, of course, does not include the numerous technical committee papers that have been published in international glass magazines.

In 1982 a reform was undertaken that aimed at closer association of the member organizations, particularly those most isolated, in an improved coordination of activities and to encourage increasing the participation of younger people. Titles were changed from then on: the Committee on Science and Technology was called the Coordinating Technical Committee (CTC) and the subcommittees were called "technical committees" (TCs).

The more formalized structure allowed a much better evaluation of the efficiency of the committees and ways to revitalize their membership. This led to the dismantling of certain subcommittees that had fulfilled their mission and the creation of others over the years.

Technical Committees

The TCs form a highly important part of the activities of the ICG. Supervised by the Coordinating Technical Committee, there are 21 active TCs composed of more than 350 members, experts in their particular fields and drawn selectively from universities, institutions, and industry in the member countries of the ICG. At present, the ICG has member organizations in 33 countries, as shown in Appendix A.

With more than half of the TC members employed in the glass industry, this structure provides an excellent vehicle for communication and exchange of information between industry and academia. The professions of TC members break out as follows: 51% work in industry, 27% in institutions, 16% in universities, and 6% in some other environment. Sixty-four percent of TC members live in Europe, 19% in the Americas, and 17% in Asia.

An important development in recent years was the combination of the TCs on publications and terminology and on documentation to form a new TC in 1996 called "Information and Communication," in line with the worldwide emphasis and importance of this field. One of the first tasks of this TC was the creation of a web site for the ICG on the Internet that contains all relevant information on ICG events and activities, including the annual reports and planned activities of the TCs. Its address is <www.icglass.org>.

The titles of the present technical committes, together with a short summary of current and recent topics, are given in Appendix B. New TCs are in preparation on glasses for medicine and biotechnology and on waste vitrification.

To promote and facilitate cooperation between TCs, clusters have been formed of committees with related objectives, coordinated by one of the CTC members. At present, clusters exist in the following fields:

- Information (TCs 1 and 23).
- Glass surfaces (TCs 16, 17, 19, and 24).
- New glasses (TCs 7 and 20).
- Glass properties (TCs 2, 3, 6, 10, and 26).
- Glass production (TCs 11, 13, 14, 15, 18, 21, 22, and 25).

From the number of TCs in the cluster on glass production, it is obvious that the ICG emphasizes topics of primary interest for the glass manufacturing industry.

A complete overview of current and recent activities of each TC can easily be obtained from the ICG website, including member lists and contact addresses of the CTC and TCs. This information can also be found in the annual report of the Coordinating Technical Committee, which is presented every year at the council meeting held during the ICG annual meetings and congresses. It may be obtained from the national member organizations through their national representatives. In addition, ICG publishes an annual report on its activities, which includes a summary of TC activities.

It should be pointed out that all publications and proceedings of the ICG and its TCs are available to the general public and are not restricted to ICG or TC members. Annual reports of the TCs are included on the ICG website and a summary of all activities is included in the ICG annual report, which is also freely available and can be obtained through the DGG.*

The activities of a TC may cover a wide range, depending on the demand in a certain area of investigation, and can certainly vary with time. The main activities, however, can be classified as follows:

- Discussions on selected topics among the members.
- Round-robin tests to establish (the accuracy and reliability of) measurement, analytical, or modeling methods.
- Development of new standards.
- Literature surveys.
- Organization of topical sessions at international conferences.
- Short training courses and workshops, mostly on technological subjects.

This shows clearly how important the role of TCs can be in the exchange of information on an international scale. All activities are done on a strictly voluntary basis and, although it is obvious that members benefit from this close contact with colleagues from other backgrounds, it is still a memorable fact that so many scientists and technologists with tight schedules and full agendas are prepared to join in these activities.

*Publications of the International Commission on Glass. For inquiries and orders, contact Verlag der Deutschen Glastechnischen Gesellschaft, Siemensstrasze 27, D-63071 Offenbach, Germany. Fax: +49-69-97 58 61-99. E-mail: <dgg@hvg-dgg.de>.

Appendix A: Organizations Participating in the International Commission on Glass (2003)

- American Ceramic Society, ACerS, USA
- Associacao Technica Brasileira das Industrias Automaticas de Vidro, ABIVIDRO, Brazil
- Balai Besar Industri Keramik, Indonesia
- Bulgarian Chemical Society
- Central Glass and Ceramic Research Institute, India
- Ceramic Society of Japan, CSJ, Japan
- Chinese Ceramic Society, CCS, People's Republic of China
- Czech Glass Society, CGS, Czech Republic
- Deutsche Glastechnische Gesellschaft, DGG, Germany
- Glasforskningsinstitutet, GLAFO, Sweden
- Glass Manufacturers Association of the Philippines Inc., GMAPI, Philippines
- Glass Manufacturers Club of the Federation of Thai Industries, Thailand
- Hellenikos Hyalourgikos Syndesmos, Greece
- Institut du Verre, IV, France
- Institute of Glass and Ceramics, Poland
- Institute Scientifique du Verre, InV, Belgium
- Instituto Nacional de Tecnologia Minera, INTEMIN, Argentina
- Institutul National de Sticla, Rumania
- Irish Glass Federation, Ireland
- Ivoclar, Liechtenstein Korean Ceramic Society, Korea
- National Comite van de Nederlandse Glasindustrie, NCNG, Netherlands
- National Commission on Glass, NCG, Russia
- National Research Centre, Egypt
- Ordem dos Engenheiros, Portugal
- Slovak Glass Society, Slovakia
- Sociedad Espanola de Ceramica y Vidrio, SECV, Spain
- Society of Glass Technology, SGT, United Kingdom
- Stazione Sperimentale del Vetro, SSV, Italy

- Szilikatipari Tudomanyos Egyesulet, Hungary
- Tamglas, Finland
- Turkiye Sise ve Cam Fabrikalari, Sise Cam, Turkey
- Vitro, Mexico

Appendix B: Recent and Current Technical Committee Topics and Activities

TC 1. Information and Communication: ICG website; ICG publications; database of committee members.

TC 2. Chemical Durability and Analysis: Publication of a book collecting all papers of the TC in the last 15 years; development and certification of reference glasses for Cr^{6+} and trace element analysis; analytical methods for determination of B_2O_3, total sulfur, and Hg in glass; measurement uncertainty of XRF analysis of glass and raw materials; certification of BCR/CRM 491 pharmaceutical glass vials.

TC 3. Basic Glass Science: Identification of standard test samples and simulation parameters for evaluation and comparison of data from different laboratories; development of standard formats for comparison of simulations and models with experiment.

TC 6. Mechanical Properties of Glass: Round-robin on edge-strength testing of flat glass; strength of "old" float glass; ways to improve edge strength; effect of edge finish.

TC 7. Nucleation, Crystallisation and Glass Ceramics: Publication of book on surface nucleation; crystallization of lithium disilicate glass ceramics; nucleation and crystallization of biomaterials.

TC 10. Optical Properties of Glass: Round-robin on emissivity measurements, especially low-emissivity coatings; measurement of optical properties of diffusing glazings.

TC 11. Contact between Glass and Refractories: Radioactivity in refractory materials; catscratches in container glass; round-robin on analysis of knots; glass-refractory contact problems (especially fused cast AZS).

TC 13. Environment: Measurement procedures of emission components; round-robin on selenium sampling and analysis; measurement of particulate and gaseous boron; implementation of European Reference Document BREF.

TC 14. Gases in Glass: Development of standard material for water in glass; round-robin on analysis of reactive gases (SO_2) in bubbles; round-robin on analysis of CO-containing bubbles.

TC 15. Sensors and Advanced Control: Defining of needs for sensors and advanced control in glass manufacturing; sensors for quality control and automatic inspection.

TC 16. Sol-gel Glasses: Publication of handbook on sol-gel technologies for glass producers and users; organization of workshops and symposia on sol-gel; special website <www.solgel.com>.

TC 17. Archaeometry of Glass: Organization of symposia during ICG congresses on ancient glass research, conservation of historical glass objects, crizzling, etc.

TC 18. Properties of Glass Forming Melts: Collection of data on glass melt properties and measurement methods; round-robin on transmission spectra at high temperatures (800–1400°C) for clear soda-lime glass and TV panel glass; round-robin on density and thermal expansion measurements for the same glasses; measurement and modeling of liquidus temperature and crystalline phases during cooling of glass melts.

TC 19. Glass Surface Diagnostics: Study of glass surfaces; corrosion of soda-lime float glass; thin films on glass (density, thickness, roughness [GIXR] [round-robin]; film thickness [round-robin]; organic layers on glass); "old" glasses (heavily corroded glass surfaces; development of standard glass for TC 17); exploration of new techniques.

TC 20. Glasses for Optoelectronics: Effect of crystallization on optoelectronic devices.

TC 21. Modelling of Glass Melting Processes: Round-robin studies on existing and hypothetical glass tanks to evaluate and compare software packages; combined glass tank and combustion space modelling; batch blanket models; sensitivity of modeling results on values of high temperature glass properties.

TC 22. Electrochemical Behaviour of Glass Melts: Validation of voltammetric methods for the characterization of redox reactions in glass melts; electrochemistry of dissolved sulfur in soda-lime glass (round-robin with standard samples).

TC 23. Education and Training in Glass Science and Engineering: Participation and organization of training courses worldwide, mainly on glass technology; bibliography of glass literature for study (web site); evaluation of education in glass science and engineering (questionnaires on available courses).

TC 24. Coatings on Glass: Test methods for mechanical properties of coated glass; methods to characterize photocatalytically active glazing (self-cleaning glass).

TC 25. Modelling of Glass Forming Processes: Evaluation of software packages for three-dimensional modeling of glass pressing processes; evaluation of software packages for modeling of the gob forming process for TV glass production; modeling of continuous fiber drawing.

TC 26. Biosolubility of Glass: In-vitro biopersistence of high-alumina, low-silica (HT-type) glass and rockwool fibers; laboratory approaches to predict the dissolution of HT fibers in animals; quantification of the extent of fiber breakage during clearing of MMVF 34 fibers.

Reference

1. *ICG 2000. History and Vision.* Edited by H. de Waal. ICG Publication, 2000.

Bibliography

R. Akçakaya, S. Isevi, and A. S. Yaraman, "International Congresses on Glass: A 65-Year Record of International Glass Research"; in *Proc. 18th Intern. Congress on Glass,* San Francisco, 1998.

Recent Developments in Chemically Strengthened Glasses

David J. Green
Department of Materials Science and Engineering, Pennsylvania State University, University Park, Pennsylvania

To open new applications or develop more efficient structural designs for glass, there is a need to identify new processes to strengthen glass. One such approach is the chemical strengthening of glass. This approach can lead to other important benefits, notably improvements in the resistance to stress corrosion and contact damage. Recently, it has been shown that engineering the shape of the surface profile produced by chemical strengthening can lead to other improvements. With these engineered stress profile (ESP) glasses, strengths can be increased while decreasing strength variability. In ESP glasses, surface cracks are arrested and this can lead to multiple cracking as a warning of failure. The phenomenon of multiple cracking implies that the surfaces of these glasses can be damaged without any loss of strength, and this has been confirmed experimentally. An overview of the processing techniques used to produce ESP glasses, the relationship of the processing to the final stress profile, the resultant mechanical properties, and the need for accurate stress measurement are reviewed.

Introduction

When glass is used in a structure, engineers usually have to use low design stresses for the glass to ensure reliability. After forming glass from the melt, it can possess very high strengths, especially in the form of fiber.[1] The problem is that glass surfaces are easily damaged by contact events, aging processes, and stress corrosion. Thus, after typical handling conditions, the strength is low and variable. The low contact damage resistance of glass is a consequence of its poor fracture toughness. This brittleness also means that failure can occur without warning — and for long-range applied stress fields, the failure is catastrophic. In extreme situations, the fragments produced by breaking glass are often a major source of injury. Clearly, it is important to identify means to strengthen glass and to control the failure process. Such a strategy opens possibilities for producing more efficient, lighter designs and developing new or safer products.

Strengthening Processes

Traditionally, thermal tempering or chemical strengthening has been used to strengthen glass products. These processes have been developed based on the observation that glasses invariably fail from surface flaws under the action of tensile stresses. The aim of the strengthening process is to place the glass surfaces in residual compression. Because residual stresses in an object must balance, these strengthening approaches place the interior of the article in tension. If the interior tensile stresses are low or if the defects in the interior are benign, failure is still expected to initiate at the glass surface, but at much higher stresses. The applied stresses must first overcome the residual surface compression before failure is possible. To a first approximation, the glass should be strengthened by the magnitude of the surface compression.

In the thermal tempering process, a glass is cooled quickly from above the glass transition temperature.[2] This process freezes a more open structure in the glass surface than in the interior. The compatibility between the surface and the interior must, however, constrain these strains. Thus, the surface region (higher volume) must be compressed by the interior of the glass, while the lower volume interior is under tension from the outer part of the glass. A similar approach is to use a process that changes the chemical composition of the surface region. Ion exchange is a well-established process for chemically strengthening glasses and it is often accomplished by exchanging Na ions in a glass with larger K ions from a melt.[3] The crowding effect produced by the larger K ions induces the surface residual compression when the exchange is performed below the glass transition temperature.

Mechanical Properties of Ion-Exchanged Glasses

Fatigue Properties

Frequently, ion exchange is optimized by identifying exchange temperatures and times that lead to maximum strength. This process can be seen as a competition between the penetration depth for the exchanging ions and stress relaxation. Ion exchange depths are usually <100 μm and in some glasses, such as soda-lime-silicate glasses, the diffusion coefficients associated with the ion exchange process can be low.[4] This leads to very long processing times to obtain the required exchange depth.

Figures 1 and 2 show dynamic fatigue data for a soda aluminosilicate (SAS) glass and a soda-lime-silicate (SLS) glass.[5] Dynamic fatigue data is often used to assess the resistance of a glass to stress corrosion because the glass strength is dependent on stressing rate if a time-dependent process is

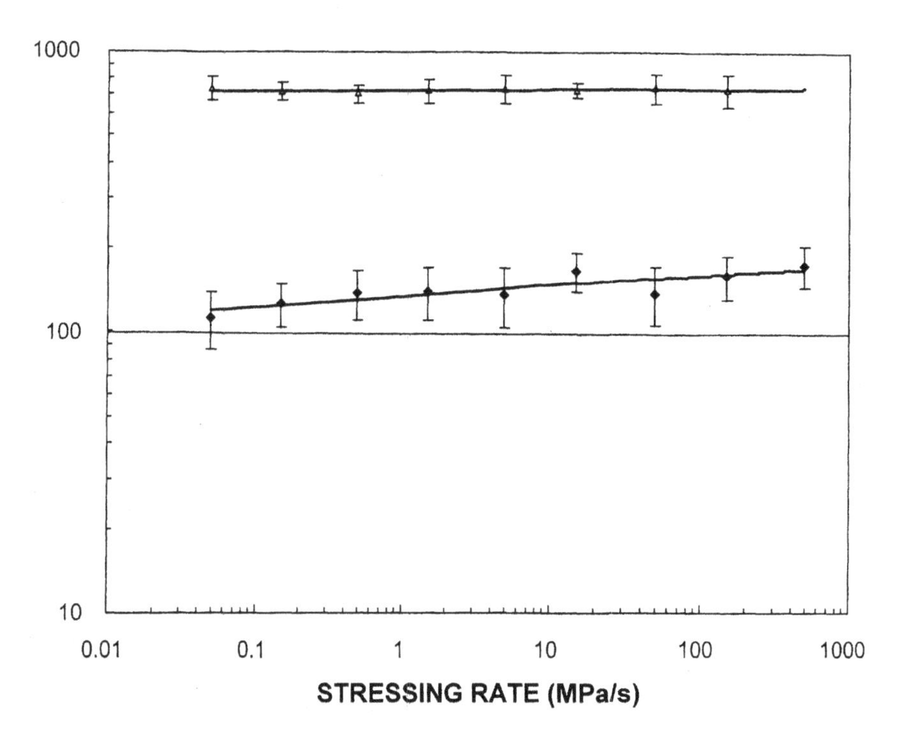

Figure 1. Strength as a function of stressing rate for an annealed and ion exchanged sodium aluminosilicate glass.[5] The subcrtical crack growth exponent for the annealed glass was 25.9 and the ion exchange conditions were 500°C (1 h). The error bars represent ±1 standard deviation.

controlling the failure.[6] As seen in the figures, the strength of the SAS glass is increased by 560–610 MPa, while the SLS glass is increased 150–200 MPa. To a first approximation, the strengthening is independent of stress rate. This is not obvious in viewing these figures because the slopes of the fitted lines are significantly different. This is the result of logarithmic axes, which make the difference between the lines look as if it is not constant. It has been suggested by previous researchers[7,8] that surface compression must be overcome before stress corrosion is possible, and this is consistent with the data shown in Figs. 1 and 2. The relationship between strength (σ_f) and stressing rate ($d\sigma/dt$) for the residually stressed glass can be expressed as

$$\sigma_f = \left[B(n+1) S^{n-2} \frac{d\sigma}{dt} \right]^{1/(n+1)} - \sigma_s \qquad (1)$$

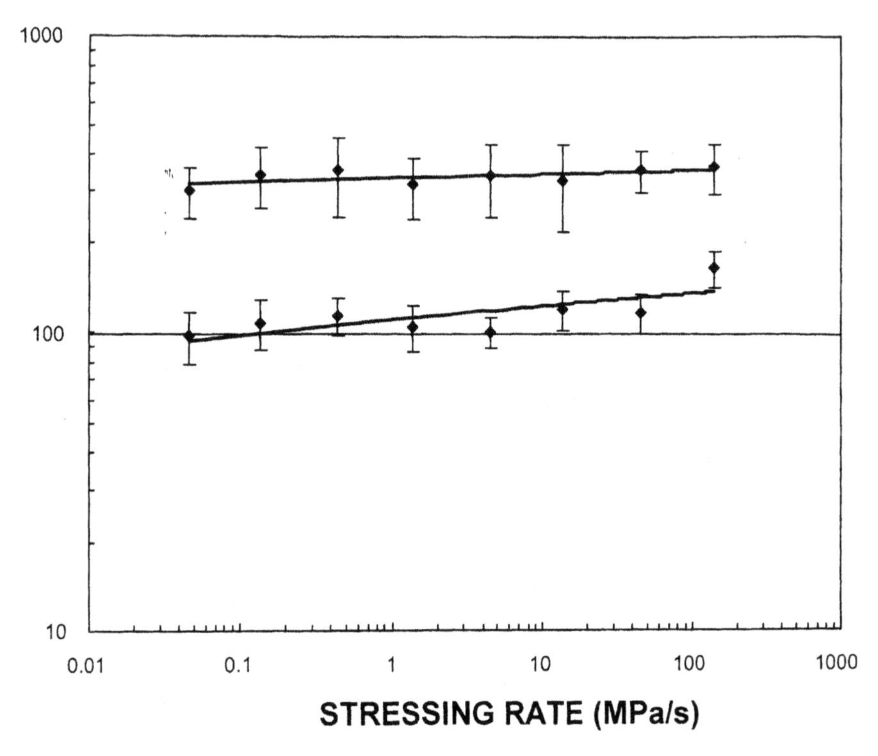

Figure 2. Strength as a function of stressing rate for an annealed and ion exchanged sodium-lime-silicate glass.[5] The subcritical crack growth exponent for the annealed glass was 21.8 and the ion exchange conditions were 450°C (1 h). The error bars represent ±1 standard deviation.

where B and n are the subcritical crack growth parameters and S is the strength of the glass prior to strengthening. Setting $\sigma_s = 0$ allows the empirical equation that is often to describe the dynamic fatigue behavior of brittle materials to be returned.[6] Equation 1 assumes the surface flaws are subjected to a uniform stress, which is questionable in ion-exchanged glass.

If the above interpretation is correct, strengthening glasses using surface compression has important consequences for design. Consider the situation when a glass is subjected to a constant stress (static fatigue), as shown schematically in Fig. 3. Some glasses possess fatigue thresholds, so this idea that the surface compression must be overcome before stress corrosion is possible implies that the fatigue threshold will also be increased by this stress difference. As shown in Fig. 3, the fatigue threshold can undergo a significant increase and becomes a much larger fraction of the glass

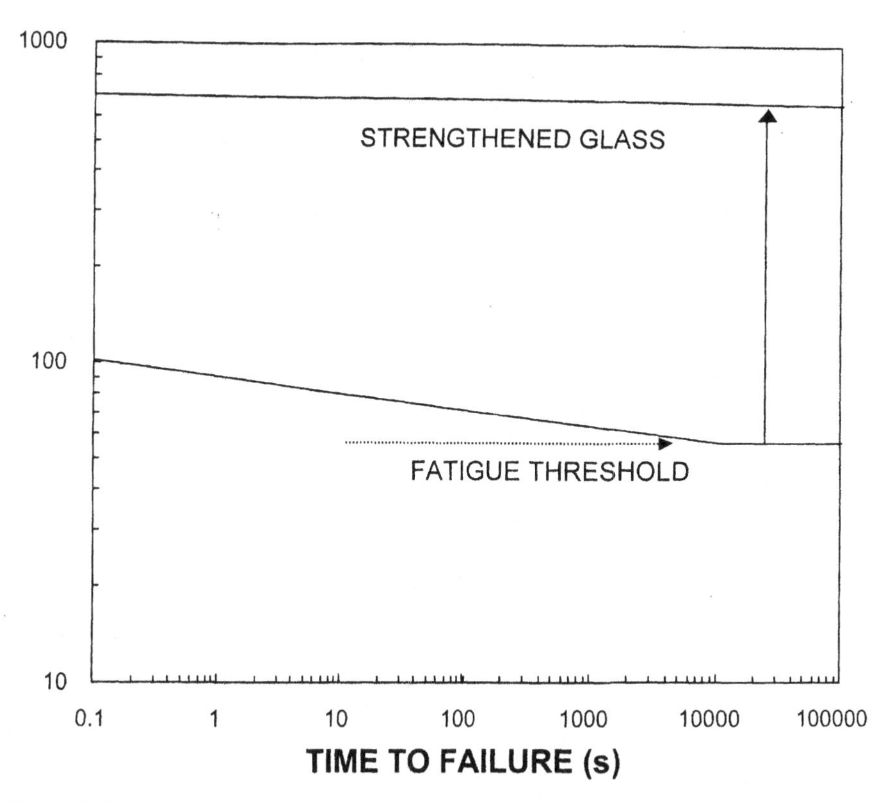

Figure 3. Schematic of change in static fatigue behavior produced by chemical strengthening. The approach assumes the stresses are increased by the magnitude of the surface compression.

strength for the strengthened glass. There is a need to confirm this behavior with future research. The main concern with this postulated fatigue behavior is that a strengthened glass can be damaged in service and, if the resultant crack exceeds the depth of the exchange, the strengthening can be partially or completely lost.

Strength Variability

Glass strength is known to be rather variable. For example, if the average strength of a glass specimen is 100 MPa, the standard deviation is typically ~20 MPa. If these glasses were subjected to uniform surface compression, the standard deviation in the strength would remain unchanged because all specimens would be strengthened by the same amount. This would lead to significant improvements in the coefficient of variation for the strengthened materials. The difficulty is that surface compression produced by ion exchange rarely leads to uniform surface compression. If the surface compression decreases with distance from the surface, it is expected that weak specimens with larger flaws will be strengthened less than those with small flaws. This behavior was explored by Tandon et al.,[9] who showed that such situations could lead to significant increases in the standard deviation of the strength. It is therefore recommended that careful attention always be given to the strength variability associated with chemical strengthening processes. For the data presented in Figs. 1 and 2, the standard deviation for the SAS was more than doubled by the ion exchange and for the SLS glass, it was more than quadrupled.

Contact Damage

As mentioned earlier, a major concern in ion-exchanged, strengthened glasses is that a large surface flaw could be produced during service that has a depth greater than that of the surface compression. This behavior can be assessed using surface indentation using a hardness indenter. For example, Tandon and Green[10] have shown that the load required to nucleate a crack at the indentation site can be significantly increased for ion-exchanged glasses. An alternative approach is to measure the strength after indentation. Figure 4 shows such data for an ion exchanged SLS glass.[11] At low indentation forces, the strength is not reduced significantly by the indentation but there is a critical force above which significant strength degradation occurs. Indeed, at very high forces, the strengthening loss can be almost complete. It is, therefore, important to determine approaches to increase the critical force and to relate this behavior to the service conditions.

Engineered Stress Profile Glasses

A new approach for chemically strengthening glass by ion exchange was introduced in 1999.[12,13] The key feature in this new approach was to carefully design the residual stress profile in such a way as to move the maximum compression away from the external surface and to control the stress gradient in the surface region. These glasses have been called ESP (engineered stress profile) glasses. ESP glasses show significant strengthening, low strength variability, and crack arrest. ESP glass surfaces undergo multiple cracking at designed stress levels and the multiple cracking acts as a warn-

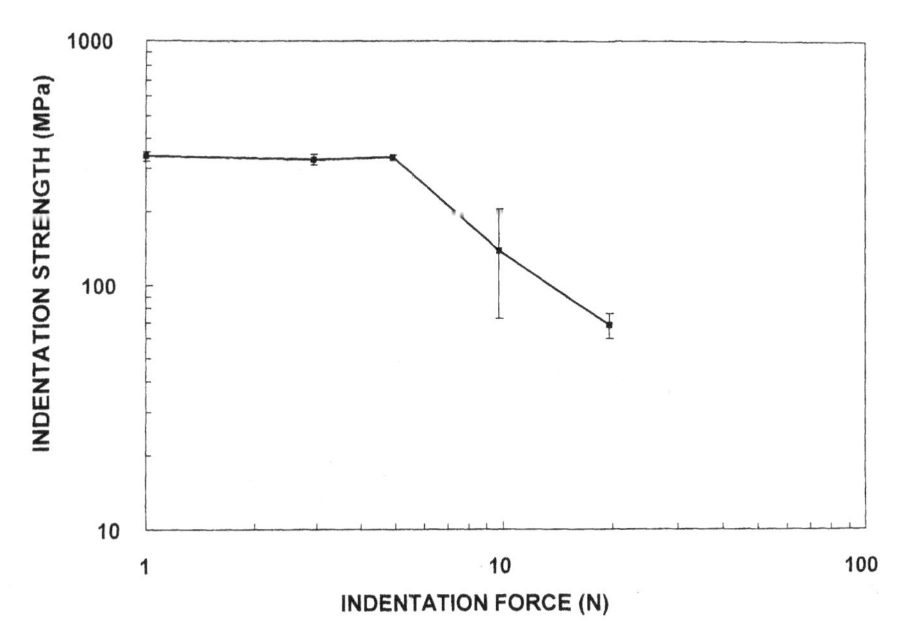

Figure 4. Indentation strength for an ion-exchanged sodium lime silicate glass. The ion exchange conditions were 450°C (48 h). The error bars represent ±1 standard deviation.

ing of the final failure that occurs at higher stress. In a sense, the shaping of the stress profile means that cracks are "trapped" within the surface region.

Processing Approaches

The first demonstration of crack stabilization in glasses using residual stresses was accomplished using a two-step ion exchange process.[12–14] The first exchange step was a typical process, in which the glass was immersed in molten KNO_3. In the second exchange, however, the glass was immersed in a mixed $NaNO_3/KNO_3$ molten salt. The idea was to remove a small fraction of the potassium introduced in the first exchange from the near-surface region. The two-step exchange process leads to control of the potassium concentration gradient and a subsurface maximum in the K concentration. In the absence of any stress relaxation, the stresses are proportional to the difference between the composition and the average composition. Thus, composition control allows the residual stress profile to be tailored. In reality, however, stress relaxation will modify the direct correlation between the

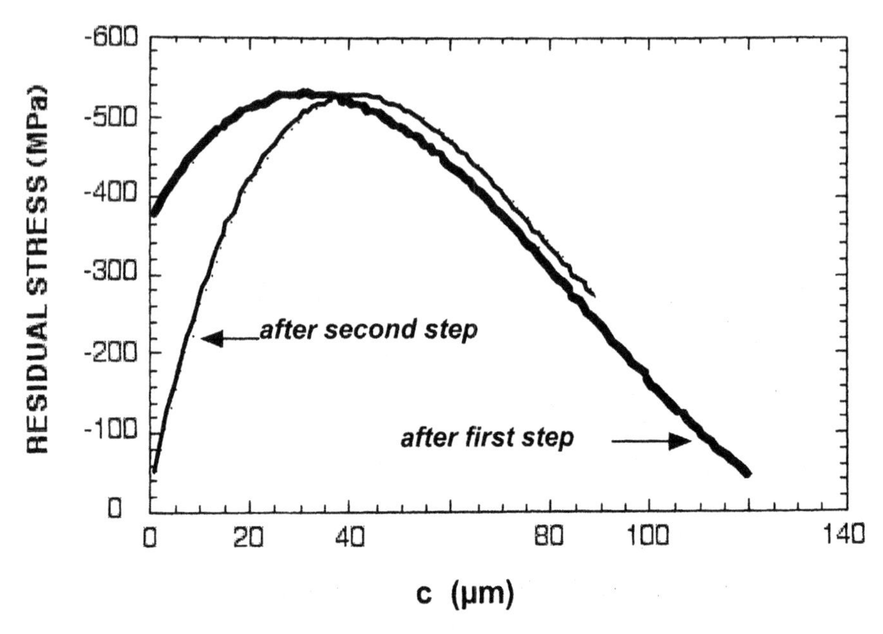

Figure 5. Comparison of stress profile produced by a traditional (single-step) process and one in which the profile is modified by a second exchange step; c = distance from surface.

composition and stress profiles. Figure 5 shows an example of the way the stress profile is modified in an SAS glass by a two-step process.

The use of mixed $NaNO_3/KNO_3$ salts in the second exchange step was important in controlling the composition gradient in the near-surface region. Indeed, if molten $NaNO_3$ was used the glasses often underwent spontaneous cracking during the processing. It was postulated that stress relaxation occurred during the first exchange and that when the Na ions were exchanged back, the surface stress would be tensile.[14] It was, however, found that these glasses with surface cracks were still stronger than the (nonstrengthened) annealed glass.

A second process has been developed to produce ESP glasses.[15] In this approach, a single KNO_3 salt bath is used but the temperature is varied during the exchange. In the initial stage, the temperature is higher than used in a normal exchange. Thus, K is introduced into the glass surface but the stress undergoes significant relaxation. The temperature is then lowered to a more typical value and the compressive stresses build up, but primarily below the surface. This variable temperature ion exchange process is expected to be more commercially attractive than the two-step process.

Stress Profile Prediction

In order to optimize the processing of ESP glasses, it would be useful to predict to predict the stress profiles directly from the processing conditions. To accomplish this methodology, one has to first predict the composition profiles from diffusivity data. These data must then be combined with the strain produced by the exchanging ions and the stress relaxation behavior.[16] Composition profiles were determined for an aluminosilicate glass subject to a single exchange (450°C) and to various second exchanges at 400°C.[17] Analysis of these data allowed the diffusivity for the Na–K exchange to be determined as a function of temperature.[17] To a first approximation, it was found the diffusivity was the same for both the first and second exchange steps and was not significantly influenced by the surface stress. This conclusion makes prediction of the composition profiles reasonably straightforward for both the two-step and the variable temperature processes that have been developed for ESP glasses.

To determine the volumetric (elastic) strain associated with the Na–K exchange, ion exchange was performed at temperatures and times for which stress relaxation could be neglected. Comparison of the stress and composition profiles then allowed the dilatation coefficient, B, for the exchange to be determined. B was found to vary with composition, decreasing from 0.066 to 0.017% per wt% K_2O as the K_2O content increases from 0 to 20 wt%.[18] Although the K ions can be initially accommodated in the glass structure as an elastic strain, another process must be available as the amount of potassium increases, reducing the dilation.

A critical aspect of the experimental methodology, outlined here, is the ability to accurately measure the stress profiles. As shown in Fig. 5, stresses can vary by several hundred MPa over distances <50 μm. In order to accomplish these measurements, a technique suggested by Beauchamp and Altherr[19] was used and further developed.[20] In this technique, the tensile stresses in the interior of the glass samples are determined from the optical birefringence of the glass. The specimens are then successively etched and the change in the interior tension and specimen dimensions can be used, via a force balance, to determine the stress that existed in the removed surface layer.

The most challenging aspect in predicting the stress profiles is to include the effect of stress relaxation. The first approach was to study the compressive stress relaxation behavior for the glass prior to ion exchange at temperatures typical for ion exchange.[21] Using viscoelastic models to fit these data, it was possible to predict the stress profile from the composition data. The difficulty is that the predicted stress values are significantly lower than the measured values.[18] More recent work has shown the stress relaxation decreases at high K content.[22] This observation indicates that the potassium in the exchanged layer significantly decreases the degree of stress relaxation.[18]

Failure Prediction Using Fracture Mechanics

The initial theoretical approach considered a single surface crack in a linear elastic material, which was subjected to a combination of residual and applied stresses. For illustration purposes, it was useful to consider residual stresses with a parabolic distribution.[23–25] Using a fracture mechanics analysis, these stress profiles can be transformed into apparent fracture toughness curves. The term "apparent" is used because the residual stresses are included in the fracture toughness of the glass. The fracture mechanics analysis allows design maps to be developed.[25] These maps allow the processing parameters needed to avoid spontaneous cracking and induce stable crack growth to be identified. The key parameters are the depth below the surface of the compression maximum (d), the distance over which the stresses fall from their maximum value to zero (λ), the fracture toughness of the glass prior to strengthening (T_0), and a parameter (T^*) that depends on the product of maximum compression magnitude and d.[25] For a given T^*, an optimum value of d/l is required. If d/l is too high, the glasses can crack spontaneously during processing. As d/l decreases, stable crack growth is predicted and the stress at the onset of this growth increases.[25] If d/l is too low, however, there is no stable growth produced by the residual stresses, and failure, once initiated, is catastrophic. The above analysis has been extended to stress profiles with a general, polynomial form.[26] This approach allows more complex stress distributions to be considered. For example, one can calculate the apparent fracture toughness curves from the measured stress profiles using a polynomial fit.

Mechanical Behavior

An important consequence of the introduction of stable crack growth is that ESP glasses can exhibit multiple cracking. Stable crack growth requires an

Figure 6. Video frame showing multiple cracking on the tensile surface of a four-point bend specimen (vertical uniaxial stress).

increase in the applied stress and, thus, after the first defect forms into crack, other cracks can form. As the stress increases, the cracks increase in density. Unloading samples prior to failure or in situ observations during loading[27] allow the multiple cracking processes to be studied (see Fig. 6). Table I shows a comparison of strength data for ESP SAS glasses with annealed glass.[14,28] For the ESP glasses, the coefficient of variation in the strength is <3% for the two-step process.

The fracture mechanics theory outlined earlier allows one to predict the multiple cracking stress and strength. To this point, the predicted and measured strengths are usually within 20%. The difficulty with the theory outlined in the last section is that it does not consider the shielding effect produced by the multiple cracking. In these cases, the strengthening will be

Table I. Comparison of strength data for sodium aluminosilicate glass

Glass	Type	Strength (MPa)
SAS	Annealed	107 ± 17
SAS	Two-step ESP	579 ± 14
SAS	Two-step ESP	515 ± 13*

*All strength measured in four-point except *; biaxial flexure.

greater than predicted. The in situ observations of the multiple cracking have also shown that cracks, after some initial growth normal to the surface, change path and run parallel to the surface. This deflection of the crack path may also have an influence on the predicted stress behavior and the crack spacing. The multiple cracks produced on the glass surface are sometimes difficult to observe after unloading, if this deflection has not occurred. Finally, it also has to be recognized that cracks deeper than the compression maximum prior to exchange can diminish or even eliminate the strengthening. The multiple cracking processes can initiate, but their extent is diminished when propagation of the larger cracks intervenes.

References

1. C. R. Kurkjian, P. K. Gupta, R. K. Brow, and N. Lower, "Intrinsic Strength and Fatigue of Oxide Glasses," *J. Non-Cryst Sol.,* **316** [1] 114–24 (2003).
2. R. Gardon, "Thermal Tempering of Glass"; in *Elasticity and Strength in Glasses.* Edited by D. R. Uhlmann and N. J. Kreidl. Glass Science and Technology, Vol. 5. Academic Press, New York, 1980.
3. R. F. Bartholomew and H. M. Garfinkel, "Chemical Tempering of Glass"; in *Elasticity and Strength in Glasses.* Edited by D. R. Uhlmann and N. J. Kreidl. Glass Science and Technology, Vol. 5. Academic Press, New York, 1980.
4. C. W. Sinton, W. C. LaCourse, and W. J. O'Connell, *Mater. Res. Bull.,* **34** [14] 2351–2359 (1999).
5. P. Dwivedi, "Crack Shape Evolution During Subcritical Growth of Cracks in Glasses," Ph.D. Thesis, Pennsylvania State University, 1994.
6. D. J. Green, *Introduction to Mechanical Properties of Ceramics.* Cambridge University Press, 1998.
7. J. E. Ritter, "Fatigue Strength of Silicate Glasses," *Phys. Chem. Glasses,* **11** [1] 16–17 (1970).
8. J. E. Ritter and M. S. Cavanagh, "Fatigue Resistance of a Lithium Alumniosilicate Glass-Ceramic," *J. Am. Ceram. Soc.,* **59**, 57–59 (1976).
9. R. Tandon, D. J. Green, and R. F. Cook, "Strength Variability in Brittle Materials with Stabilizing and Destabilizing Resistance Fields," *Acta Metall.,* **41** [2] 399–408 (1993).
10. R. Tandon and D. J. Green, "Indentation Behavior of Ion-Exchanged Glass," *J. Am. Ceram. Soc.,* **73** [4] 970–977 (1990).

11. M. B. Abrams, D. J. Green, and S. J. Glass, "Fracture Behavior of Engineered Stress Profile Soda Lime Silicate Glass, Glass Technology," *J. Non-Cryst Sol.,* **321** [1–2] 10–19 (2003).
12. D. J. Green, R. Tandon, and V. M. Sglavo, "Crack Arrest and Multiple Cracking in Glass Using Designed Residual Stress Profiles," *Science,* **283**, 1295–1297 (1999).
13. D. J. Green, R. Tandon, and V. M. Sglavo, "Strengthening, Crack Arrest, and Multiple Cracking in Brittle Materials Using Residual Stress," U.S. Patent No. 6 516 634, Febru ary 2003.
14. V. M. Sglavo and D. J. Green, "Flaw Insensitive Ion-Exchanged Glass: II Experimental Aspects," *J. Am. Ceram. Soc.,* **84** [8] 1832–1838 (2001).
15. J. Shen and D. J. Green, "Variable Temperature Ion-Exchanged ESP Glasses," *J. Am. Ceram. Soc.,* **86** [11] 1979–1981 (2003).
16. A. Y. Sane and A. R. Cooper, "Anomalous Stress Profiles in Ion-Exchanged Glass," *J. Am. Ceram. Soc.,* **70** [2] 86–89 (1987).
17. J. Shen, D. J. Green, and C. G. Pantano, "Composition Profiles in Two-Step Ion-Exchanged Glasses," *Phys. Chem. Glasses,* **44** [4] 284–292 (2003).
18. J. Shen and D. J. Green, "Prediction of Stress Profiles in Ion-Exchanged Glasses," *J. Non-Cryst Sol.,* submitted November 2003.
19. E. K. Beauchamp and R. H. Altherr, "Stress Determination in Opaque Materials," *J. Am. Ceram. Soc.,* **54** [2] 103–105 (1971).
20. M. B. Abrams, J. Shen, and D. J. Green, "Residual Stress Measurements in Ion-Exchanged Glasses by an Optical Method," *J. Testing Eval.,* accepted December 2003.
21. J. Shen, D. J. Green, R. E. Tressler, and D. L. Shelleman, "Stress Relaxation of a Soda Lime Silicate Glass below the Glass Transition Temperature," *J. Non-Cryst Sol.,* **324** [3] 277–288 (2003).
22. J. Shen and D. J. Green, "Effect of the K/Na Ratio in Mixed-Alkali Lime Silicate Glasses on the Rheological and Physical Properties," *J. Non-Cryst Sol.,* submitted September 2003.
23. R. Tandon and D. J. Green, "Crack Stabilization under the Influence of Residual Compressive Stress," *J. Am. Ceram. Soc.,* **74** [8] 1981–1986 (1991).
24. R. Tandon and D. J. Green, "The Effect of Crack Growth Stability Induced by Residual Compressive Stresses on Strength Variability," *J. Mater. Res.,* **7** [3] 765–771 (1992).
25. D. J. Green, "Critical Parameters in the Processing of Engineered Stress Profile Glasses," *J. Non-Cryst. Sol.,* **316**, 35–41 (2003).
26. V. M. Sglavo, L. Larentis, and D. J. Green, "Flaw Insensitive Ion-Exchanged Glass: I Theoretical Aspects," *J. Am. Ceram. Soc.,* **84** [8] 1827–1831 (2001).
27. V. M. Sglavo and D. J. Green, "Fractography of ESP (Engineered Stress Profile) Glass"; presented at the Annual Meeting of the American Ceramic Society, April 2002 (Paper AMD.3-A-06-2002).
28. D. J. Green, V. M. Sglavo, E. K. Beauchamp, and S. J. Glass, "Designing Residual Stress Profiles to Produce Flaw-Tolerant Glass"; pp. 99–105 in *Fracture Mechanics of Ceramics,* Vol. 13. Edited by R. C. Bradt et al. Kluwer Press, Netherlands, 2002.

Glass Art and Glass Science: A Mutually Beneficial Exchange

Margaret Rasmussen
Paul Vickers Gardner Glass Center, Alfred, New York

Michael Greenman and John Brown
Glass Manufacturing Industry Council, Westerville, Ohio

Over the centuries, glass artists and glass technologists have shared creative ideas and technical secrets. At one point in history, the glass artist and the glass scientist were one and the same. Many commercial glass processes have evolved from the techniques developed by artists. Likewise, ever more fascinating and beautiful glass art forms have been created by glass artists who benefited from the technical knowledge of the commercial glassmaker. Could this mutually beneficial exchange be revived to stimulate renewed excellence in the U.S. glass industry production and glass art creation? This paper will discuss a project proposed by the Paul Vickers Gardner Glass Center at Alfred University and the Glass Manufacturing Industry Council to bring together leading proponents of the glass industry and glass artists to explore ideas and exchange technical knowledge for the benefit of both the glass industry and the glass art community.

Introduction

A snapshot of the U.S. glass industry on 13 October 2003 depicts a less than dynamic picture. Research funding for glass materials has shrunk nationwide to overall less than 1% of industry sales. Profit margins have shrunk to levels that make CEOs and corporate board members uncomfortable. Manufacturing plants and equipment are aging and investors are not eager to provide capital for rebuilds. Imported glass products flood our U.S. consumer market. Other materials mimic the properties of glass adequately but at a lower cost. Should these trends continue, what will the picture of the U.S. glass industry look like in 30, 20, or even 10 years?

The U.S. glass industry has distinguished itself throughout history with countless innovations and technical advances. We invented and developed glass mold technology and exported it to Europe. We developed optical fiber applications. We advanced the science of nuclear waste vitrification and lead in biological applications. Our list of major contributions to glass science could go on.

Figure 1. Crystal Palace exhibition.

History of Glass Art/Science Exchange

Yet today our industry seems stagnant and in need of new ideas, new applications for technology, new products, new uses for advanced glasses, new methods of processing, and new ways to use glassmaking equipment. When the glass industry of England found itself in a similar state in the mid-nineteenth century, Prince Albert, Queen Victoria's German consort, devised a master plan to upgrade the English industrial arts. His brainstorm, the Crystal Palace Exhibition of 1851, revitalized England's industry, which lagged seriously behind that of other European countries.

Of all the industries, glass received the largest boost when horticulturist Joseph Paxton designed the main exhibition hall (Fig. 1) to be constructed of 300 000 hand-blown panes of glass — one-third of England's annual glass production. The entire glass industry was challenged by Paxton's specification that the cylinder-blown glass was to measure one foot longer than glass that had been made before.

This first world's fair did much to re-energize industry throughout England. Museums were set up to exhibit examples of classical design for study and to encourage appreciation for the commercial art produced in English factories. Collections were imported from China, Japan, and the Far East as examples to inspire industrial workers in artistic design and in science. The government established schools throughout the English districts to educate English consumers to appreciate the wares of English factories. Men were educated in science and art. Women were trained as artists to work in factories rather than limited to being governesses or servants. Industrial workers were taught art and science in equal portions.

At this time in glassmaking history, the glass artist and the glass technologist were one and the same. The artist smock doubled as a lab coat. Glass was made by the team method, and glassmakers especially enjoyed a status in society of labor elite. Glass art and glass science enjoyed a mutually beneficial exchange, beginning with education as the key. The result was excellence in glassmaking.

The benefits of the symbiotic relationship between art and science become evident with an exploration of early twentieth century glass. From design of batch components to forming methods and finishing techniques, the roots of science and art were intertwined in the glassmaking process.

Science Learns from Art

From 1905 into the 1920s, the artist-scientist Frederick Carder, founder of Steuben Glassworks in Corning, New York, added silver oxide to clear glass batch to create an iridescent glass he called "Aurene." When a glass object was blown and nearly formed, Carder put it into a reducing atmosphere that drew the silver oxide to the surface of the glass and formed a mirrorlike skin. When the surface was sprayed with stannous (tin) chloride, the object turned iridescent blue or gold, depending on the thickness of the tin coating (Fig. 2).

In 1928 Corning scientist Jesse Littleton was exploring ways to eliminate radio static caused by the buildup of electricity in the glass insulators on utility poles. He measured the surface resistance of one of Carder's Steuben iridescent glasses and discovered that it conducted electricity. By

Figure 2. Aurene.

Figure 3. Corning Pyrex insulator.

using Carder's process to create a tin oxide coating on the insulators, Littleton solved the irksome problem of buzzing and squawking in early radio transmission — a mutually beneficial partnership in art and science.

In further development of this surface technology, doping of tin oxide with antimony by Mochel in 1945 led to precision electronic resistors, heat reflective coatings on glass and on everything from home and office heaters to aircraft windshields. Low-E energy-efficient window coatings at Pilkington and liquid crystal displays for flat panel color TV and computer monitors at Corning Incorporated also evolved from this technology.

The batch composition of opalized or vaseline art glass produced commercially in the early 1900s contained phosphorous from bone ash, calcium, and silica (Fig. 4). The phosphorous caused the glass to begin to crystalize, or phase separate, causing the glass to opalize when a jet of compressed air was applied to a hot, formed glass object.

Scientists in glass labs today are developing human bone repair, dental implants, and spinal vertebrae spacers from glass made of the same batch components — phosphorous-calcium-silica — as opal or vaseline art glass. When the bioactive glass (Fig. 5) is heat treated, small crystals form within the glass, causing opalization or opacity. By adding a bit of lime to the batch, scientists can control the dissolution rate of the glass, so a second operation to remove a metal plate is unnecessary.

The lovely purple color of Wisteria art glass (Fig. 6) fluoresces with changes in light source due to the presence of neodymium as a component to the glass batch.

Figure 4. Vaseline/opal glass.

Figure 5. Biotech glass.

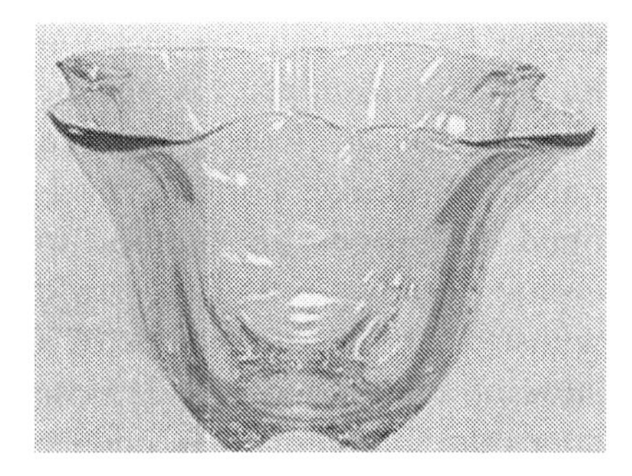

Figure 6. Wisteria glass.

Figure 7. Phosphate laser glass.

Schott Glasworks and Hoya have developed high-purity, phosphate laser glass (Fig. 7), adding neodymium to the batch of optical glass. This glass has led to development of high-energy and high peak power solid-state laser systems for fusion energy research. The National Ignition Facility, under construction at Lawrence Livermore National Laboratory, is dependent on this purple-colored wisteria glass to achieve controlled thermonuclear (fusion) ignition in a laboratory setting.

The opaque glass that Frederick Carder developed in 1916 was named "Tyrian" for the ancient Phoenician Kingdom of Tyre, which produced the rich purple dye for royal clothing (Fig. 8). A secret ingredient in the batch caused microscopic crystals to form in the glass when the temperature was high enough to cause the reaction but low enough that the object kept its shape.

Figure 8. Tyrian glass.

Chemist Donald Stookey accidentally overheated a test furnace in 1954 and unwittingly caused microscopic crystals to form and change the glass into glass-ceramic. When he dropped his opaque glass on the lab floor, it bounced rather than broke. Corning developed the processing method to create cooking pots that could withstand the thermal shock of a stovetop burner and missile nose cones durable enough to endure the harsh conditions of outer space (Fig. 9).

Figure 9. Glass-ceramic nose cones.

Perhaps the most interesting transfer comes from the ancient art known as millefiori or "a thousand flowers" (Fig. 10). Roman glassmakers made fused mosaic, or millefiori, glass as early as 1500 B.C. Venetian glassmakers carried the forming process to a greater degree of sophistication around A.D. 1500. But it was not until the 1840s that French glassmaker Jacopo Franchini revived the fusing process to any notable degree when he created half-inch, miniature portraits of royal patrons complete with dates and wording. To create art glass millefiori or mosaic glass, rods of different colored glasses with matching thermal expansion coefficients were bundled into a design and heated to fuse them

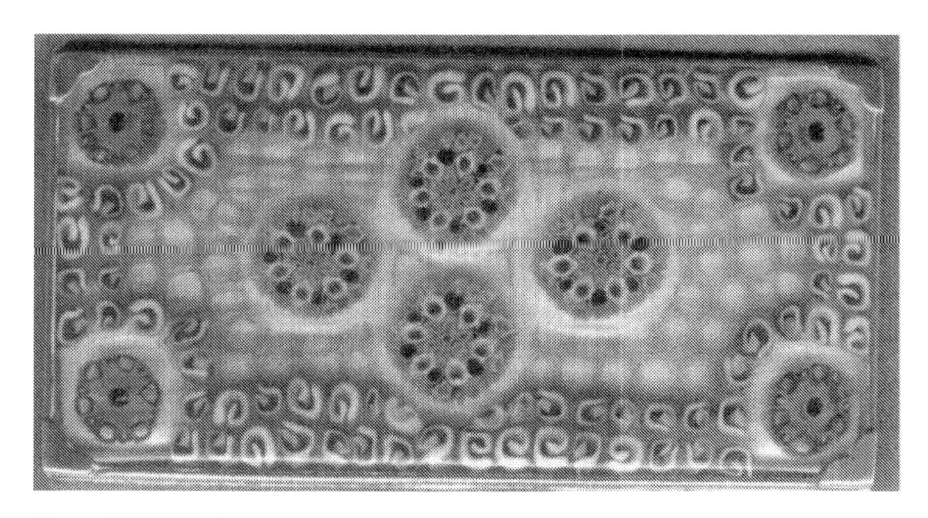

Figure 10. Millefiori.

together. One worker attached a glass gob on an iron pole to one end of the bundled rods while another worker attached a glass gob to the other end of the bundle. The second worker, usually a young boy, ran away, pulling the glass into a long cane, miniaturizing the mosaic design in the bundle while maintaining the design of the colored rods in the original bundle.

Figure 11. Optical fiber.

Bundling and fusing glass rods, miniaturizing the bundles, maintaining the integrity of the rod — does this forming process sound familiar? The process of bundling and miniaturizing for millefiori art glass has been transferred to a highly sophisticated technology to create optical fiber endoscopes. Bundles of highly pure glass rods are fused together and drawn out from fiber-drawing towers that replace the young runner. The fibers retain their integrity to convey light waves through the miniaturized bundles of fibers. The concept of guiding light through a fine glass fiber was patented in 1920 by American electrical engineer and inventor Clarence W. Hansell. The first image was transmitted through a bundle of glass fibers in 1930 by German medical student Heinrich Lamm. With the commercial development of a single cladded fiber of ultra-pure optical glass the thickness of a hair, society has been catapulted into the Information Age (Fig. 11).

Figure 12. Photochromic glass.

Art Learns from Science

The exchange goes both ways. Advances in technical glass science and engineering have added new dimensions to architectural glass art, sculpture, and all forms of art glass.

Photochromic glasses have added a whole new dimension to architectural glass art. Working in large-scale formats, glass artists can transform the profile of a skyscraper, as did Ken von Roenn with his massive sculpture atop an insurance building in Charlotte, North Carolina (Fig. 12).

Glass artists who make public artwork are concerned about strength in glass and have a high awareness of public safety. Understanding and being able to apply high-tech methods of strengthening glass such as ion exchange and advanced coatings developed for science are invaluable to any glass artist.

Renowned glass artist Dale Chihuly sought the expertise of glass scientists to successfully transfer his free-form paintings to glass surfaces. As a result, he moved into a new form for his glass art (Fig. 13).

With a scientific laser system driven by sophisticated computer software, three-dimensional images can be created within a solid block of crystal without creating flaws in the surface of the glass (Fig. 14). Laser beams penetrate the crystal and create minuscule dots within the crystal to create any desired design. Hundreds of thousands to millions of laser pinpoints are fired directly into the block of crystal glass. When beams of light refract from the design within the glass, the design appears to be a three-dimensional sculpture within the block of crystal.

And how pleased artists would be to incorporate the technology of self-cleaning glass. Using environmentally friendly, high-volume production

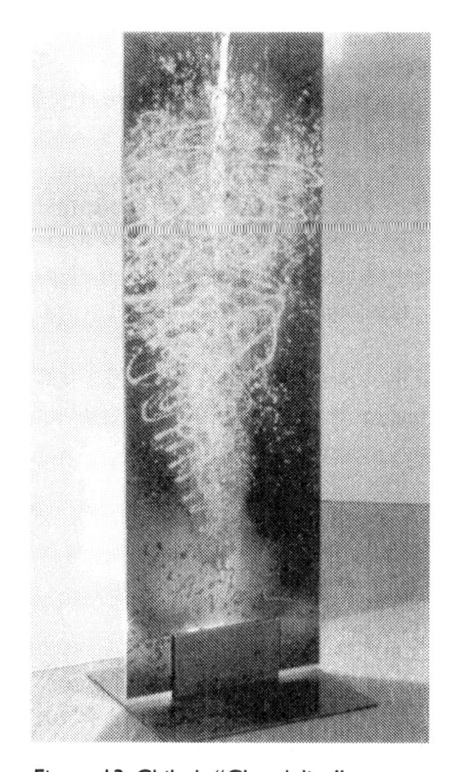

Figure 13. Chihuly "Chandelier."

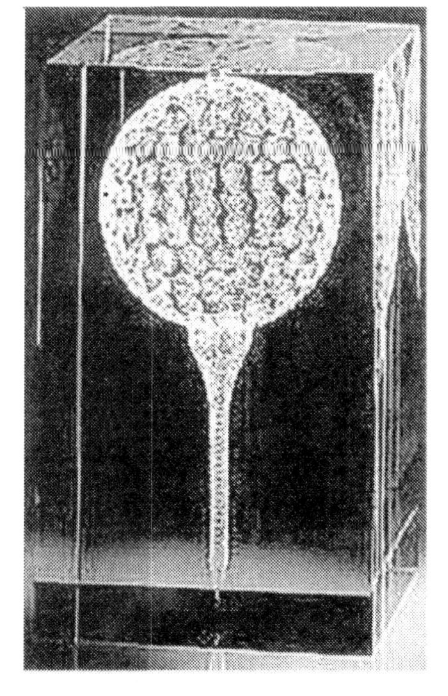

Figure 14. Laser glass art.

methods, such as roll coating, screen printing, electrostatic methods, and spraying, glass scientists at Ferro Corporation have mimicked nature's ingenious water- and dirt-repellent lotus leaf. PPG Industries has introduced self-cleaning windows based on the photocatalytic and photo-induced hydrophilic properties of titanium dioxide film.

Artistic/Scientific Exchange

But opportunities for these fruitful exchanges of creativity and technology to occur on a systematic basis are all too infrequent today. Were the process to be formalized for artists and scientists to enjoy a mutually beneficial exchange, could the U.S. glass industry be revitalized?

Take another look at the snapshot of the U.S. glass industry. What can we do to revitalize one of our country's most important industries? Might not a reunion of glass art with glass science do much to re-energize this stagnant industry? Could we not recreate the mutually beneficial exchange between glass art and glass science to develop new glass forms, new uses for traditional glasses, new applications for advanced glasses, or new ways of using color in glass products?

Might this be a solution that would reverse the equation of selling more but making less profit, as is the case in the industry today? And were that profit to be increased, would we not then have increased funding for glass research? Those in the know say we are only scratching the surface in glass research compared to the mysteries of glass science that continue to go unsolved.

By envisioning and enacting a strategy to reunite glass art and glass science, could we not increase the profit margin to the desired 35% at plant level and increase funding for needed glass science research? The transfer of art to science and science to art has occurred only coincidentally over the past century. We like to think there is a way to formalize the process of glass artists and glass scientists working in tandem for mutual benefit.

Intentional Exchanges

On a very minor scale, there has been some experimentation with the concept in an academic classroom. Alfred University glass scientists Bill Lacourse and Suhas Bhandarkar agreed to introduce use of sol-gel glasses to members of art professor Jackie Pancari's glass casting class. The scientists suggested that monolithic sol-gel forms could be colored, sculptured, spot fired, or otherwise manipulated. Students were instructed to design molds to specifications for mold coatings and materials. No restrictions were placed on their design. The student molds met with mixed success. The tendency is for the sol-gel glass to dry too rapidly and split.

One student solved this problem by using a natural sea sponge as his mold. The sponge absorbed the liquid sol-gel mixture and when fired maintained the details of the sponge.

Another student came in with a used tea bag in a disposable cup. Compared to the intricate mold designs of other students, this student seemed destined for an F. But when she slowly dipped her tea bag into the sol-gel mixture, she created a shape that did not crack during drying and had all the intricate shape that lent it to design.

As the molds dried, the roles of scientist and artist reversed. The glass scientists began to study the exotic shapes cast in balloons, wax molds, plastic, toothpicks, glass beads — molds a scientist would never have dreamed of trying. Even after the class ended, the scientists and artists continue to work together to develop applications for advanced sol-gel glass. Incidentally, Professor Bhandarkar was one of the lead scientists in development of the sol-gel formula for monolithic shapes in industry.

GMIC-PVGC Collaboration

Because we believe that the reunion of glass art with glass science will reinvigorate the U.S. glass industry economically and aesthetically, the Glass Manufacturing Industry Council in collaboration with the Paul Vickers Gardner Glass Center at Alfred University are undertaking a nationwide effort to formalize collaboration between glass artists and scientists, beginning with this paper. Our step-by-step strategic plan to formalize collaboration between glass artists and scientists begins with a series of presentations. We are introducing the concept of reuniting art and science in commercial glassmaking at major glass gatherings.

For the second step, we plan to conduct a brainstorming workshop that will bring together leading scientists and accomplished artists who can identify practical ways to formalize the relationship. They will explore ways to develop a distinctive American glass, incorporating the best of technology and the best of art glass.

The third step will be to sponsor a Working Partners Summer Camp to give glass scientists and artists an opportunity for total immersion to explore a particular problem or project of interest to a two-person team.

The fourth step will be to solicit cooperation from chief technical directors in corporations and manufacturing facilities across the country to provide experimental venues for artists and scientists to collaborate on the job and attempt to institutionalize the process.

Challenges

To move forward with this concept to revitalize the nation's glass industry, we must be realistic. We understand the challenges.

As in any corporate industry, chief executive officers and boards of directors are most concerned with quarterly earnings. Looking beyond to stability of their companies in 5–10 years is not an exercise in which they often engage, nor do they listen as eagerly to their chief technical officers as readily as they do their financial officers. CEOs usually do not have the luxury to take risks that might affect their Wall Street ratings. This is especially true in the glass industry, where flat glass furnaces go online for 10–15 year time span and capital investment requires millions of dollars.

Technology has advanced to such a sophisticated degree that scientists speak of excitation of glass at the atomic level, and artists understand molten glass as it yields to their own breath through a blowpipe or responds to their physical manipulation at the bench.

Timetables for development of new products may need to be extended for the artist and the scientist to reunite in a creative and productive process. Appropriate space for artists and scientists to collaborate in a meaningful setting will not be easy to find with proprietary interests of glass manufacturers being such an important issue.

Corporate administrators may have doubts about their scientists spending seemingly nonproductive time exploring ideas. Chief technical officers must be sold on the feasibility of the program so they can convince their CEOs and board members to share the vision.

Union rules that prohibit crossover between jobs will have to be circumvented, possibly through weekend classes for art and science, not unlike the classes conducted in late nineteenth-century England in which Frederick Carder was instructor and student.

Funding for this major project might be difficult to achieve because of the time lag between creation and application. But with government agencies collaborating with public and private corporations and academic institutions, the funding can be a cooperative venture that does not drain any one source too extensively.

Benefits

But the benefits will overshadow any challenges that we might face. We believe that a formalized reunion of science and art will reinvigorate the stagnant glass industry to yield increased financial gain and claim a higher percentage of the global market for American glass products.

Manufacturing capabilities will be increased, new products that are not commodities will be developed, innovative uses of equipment will be devised, and the quality of existing glass products will be improved for the consumer market.

Profit margins will be increased to allow a higher percentage of funding for research — still a critical need of the industry.

What do we want you to do? Show this paper to your "outside the box" glass technologists and CTOs. Discuss the concept and, if possible, obtain an agreement to participate in our brainstorming ventures. That will be our starting point to launch the full program.

American glass can achieve, in the words of one philosopher of science, Rainer Maria Rilke, a "conflagration of clarity." That is, visionaries of the U.S. glass industry will allow ingenious scientists and accomplished artists to see what none before them has ever seen, to the economic and spiritual benefit of all.

Acknowledgments

The authors would like to acknowledge the contributions of the following people: Carole Onoda, researcher, Paul Vickers Gardner Glass Center; George Onoda, consultant; L. David Pye, glass scientist; Alexis Clare, glass scientist; William C. LaCourse, glass scientist; Suhas Bhandarkar, materials scientist; and Jacqueline Pancari, studio glass artist.

Bibliography

Corning Museum of Glass, *Innovations in Glass.* Corning, New York, 1999.

P. V. Gardner, *The Glass of Frederick Carder.* Crown Publishers, New York, 1971.

J. Hecht, *Understanding Fiber Optics,* 4th ed. Prentice Hall, Columbus, Ohio, 2002.

L. L. Hench and J. Wilson, *An Introduction to Bioceramics.* World Scientific, New Jersey, 1993.

Rainer Maria Rilke, *Art & Physics: Parallel Vision in Space, Time and Light.* Leonard Shalain, Quill, William Morrow and Co., New York, 1991.

W. Vogel, *Chemistry of Glass.* American Ceramic Society, Westerville, Ohio, 1985.

Self-Repair of Glass and Polymers

Carolyn Dry
School of Architecture, University of Illinois, Champaign, Illinois

Service life of glass composites is determined by fracture and corrosion. Delaminations and cracking are the major cause of structural failure. Development of fractures and catastrophic failure can occur on macroscopic and microscopic levels at one time, or in sequence. Safety glass as a composite is composed of two materials: the glass and the reinforcing fibers. One of the main common failures occurs at the interface between the matrix and the fibers. To prevent interlaminar failure and fatigue, good bonding between the matrix and the fibers is needed. Internal damage is also common in glass, especially corrosion damage. Repair of damage is important so that failures do not progress to the ultimate catastrophic failure of breakage. But corrosion damage is hard to detect unless it has developed to macroscopic scale flaws. Nondestructive evaluation techniques have limited our ability to detect corrosion damage.

In order to be self-repairing, a healing chemical is stored in hollow fiber vessels embedded in the matrix or between layers. When the glass is damaged, the corrosion or damage progresses, opening the repair fiber and releasing the repair chemical. The healing chemical flows into the damaged area and repair takes place.

This research concerns the repair of damage by the release of chemicals from fibers into matrix corrosion and microdamage sites so that development of further damage can be prevented. The repair fibers can also release chemicals between delaminating layers or rebond fibers to the matrix. Two different repair mechanisms for self-repair have been investigated, and experiments assessed the ability to both rebond fibers and repair cracks using fiber pullout, impact, and bending tests with promising results

Impact tests on polymer specimens have revealed repair in less than 10 s. An Epon epoxy impact sample containing two-part epoxy in glass pipettes was subjected to impact in a Dynatup machine. The results were dramatic. Within seconds the two-part adhesives had filled all contiguous cracks even though no fibers had been directly hit, and the adhesives raced around and filled the circular crack, an artifact of testing caused by the edge of the impact machine itself. Adhesive flow, setup, and repair was accomplished in less than 10 s.

Self Repair: The Answer to Damage in Polymers and Glass

This research proposed that repair of microdamage be done by the release of chemicals from fibers or beads into matrix cracks so that development of further damage can be prevented. The repair vessels also release chemicals between delaminating layers or rebond fibers to the matrix. Two different repair mechanisms for self-repair have been investigated, and experiments assessed the ability to both rebond fibers and repair cracks using fiber pullout, impact, and bending tests with promising results.

In order to be self-repairing, a healing chemical is stored in hollow vessels such as fibers or beads embedded in the polymer matrix. When the composite cracks, the crack progresses, cracking the repair fiber (see Fig. 1). The healing chemical flows into the crack and the crack faces are rebonded. Also, the fiber can be rebonded to the matrix.

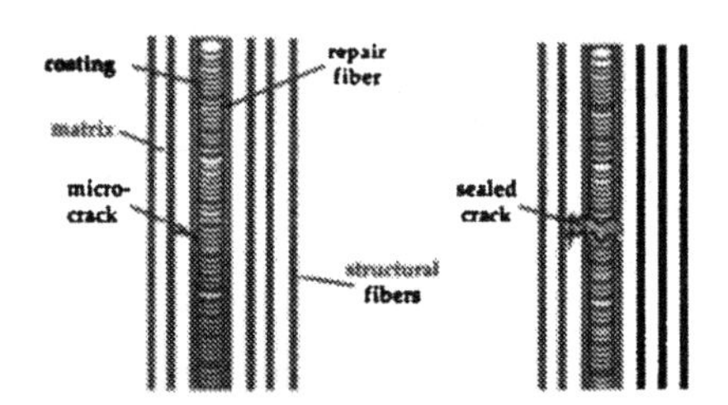

Figure 1. Schematic of repair mechanism.[1]

Impact tests on specimens have revealed repair in less than 10 s. As seen in Fig. 2, an Epon epoxy impact sample containing two-part epoxy in glass pipettes was subjected to impact in a Dynatup machine. The results were dramatic. Within seconds the two-part adhesives had filled all contiguous cracks even though no fibers had been directly hit, and the adhesives raced around and filled the circular crack, an artifact of testing. Adhesive flow, setup, and repair were accomplished in less than 10 s.

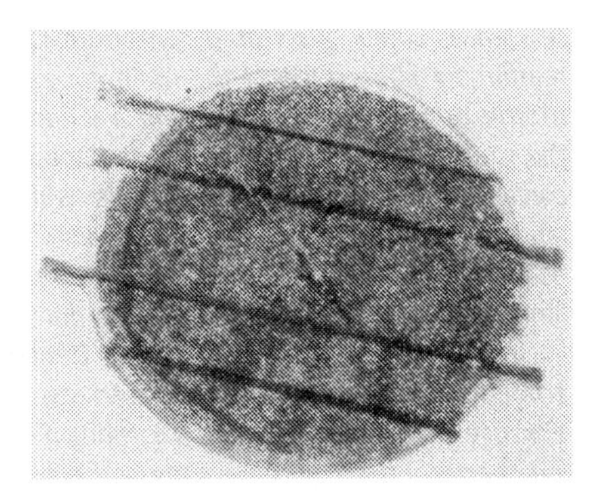

Figure 2. Photo of a 3-in. diameter impact epoxy matrix sample with double lumen glass fibers containing two part epoxy resin, dyed red and black. Gray part contains 500-μm glass spheres.[1]

Factors in the Design of the Self-Repairing Systems

Materials capable of self-repair consist of several parts: (1) an agent of internal deterioration, such as dynamic loading, that induces cracking, (2) a stimulus to release the repairing chemical, (3) a fiber or bead, (4) a coating or fiber wall that can be removed or changed in response to the stimulus, (5) a chemical carried inside the fiber, and (6) a method of hardening the chemical in the matrix in the case of cross-linking polymers or monomers.[2,3]

Further, these materials capable of passive, smart self-repair must meet the following criteria:

1. They must not degrade the matrix properties.
2. They must contain enough chemical to repair the damage.
3. The encapsulator breaks in response to damage.
4. The chemical must flow out of encapsulator.
5. Enough chemical must be released to reach cracks.
6. Crack damage must be repaired.
7. Repair must be rapid enough for the end use.
8. Overall catastrophic failure must be prevented.
9. Overall strength must be restored to 80–100% or more.
10. New cracks should form before repaired cracks reopen.
11. Additional chemical must be able to be supplied.
12. Damage and subsequent repair must be assessed.

The chemicals chosen must also withstand ambient heat extremes, have a long shelf life, be of reasonable cost, not be environmentally hazardous or have a noxious odor, and survive the heat/pressure of polymer processing.

The Influence of End Use Application on Factor Choices

One of the topics to investigate is the influence of the end use of the composite on factor selection. The assumption is that fibers were the repair vessel configuration chosen.

Sudden impacts need extremely fast response so that the first damage is repaired before the whole composite fails. Other release mechanisms approaching the speed of millisecond deployment are possible.

In order to resist high temperatures, pressure, and bending configurations, special designs are to be used in composite processing.

Additional repair chemical may need to be embedded over time for later self-repair. This can be done with fibers that are exposed at the composite edge (see Fig. 3).

Figure 3. Demonstration of the vacuum concept of refilling embedded repair fibers.

Fibers can be examined to reveal any cracks in the matrix. Fibers that extend to the edge can reveal loss of repair chemical and thus where cracks exist and the volume of cracking.

Our Research on Self-Healing Polymer Composites

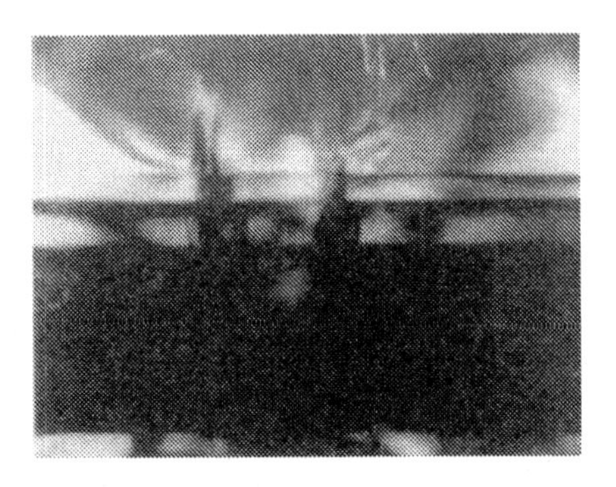

Figure 4. Microscope photo of release of an adhesive into a polymer matrix.[1]

Our investigation involved the development of polymer matrix composites that have the ability to self-repair internal cracks due to mechanical loading.[1] Most of it focused on the cracking of hollow repair fibers dispersed in a matrix and the subsequent release of repair chemicals that result in the sealing of matrix cracks, the restoration of matrix strength, and the ability to retard crack propagation. It was found that cracking of the repair fiber and release of repair chemical could be achieved. This was verified by visual and auditory inspection. Figure 4 shows the release into a polymer matrix, under microscopic magnification. Impact, fracture, and bend tests were performed and revealed the ability of this system to fill and repair cracks, restore strength, and retard crack growth. Fiber pullout tests were done that revealed that fibers were rebonded.

In this research, bending tests were done on samples containing both filled and empty repair fibers. Two different mechanisms for self-repair were investigated: release of single polymer and release of a cross-linking two-part polymer. The single polymer used was cyanoacrylate; the two-part polymer used was a two-part epoxy. In the two-part system, filled fibers were matched, one containing a monomer, the other containing a cross-linking chemical. The samples were metal fiber–reinforced epoxy matrix composite. Two layers contained three hollow glass fibers and three metal wires running the length of the samples. Both layers were below the midpoint of the sample's section. The sample was loaded in tension, causing the repair fibers to break and release their chemicals into the matrix. After sufficient time for curing of these chemicals, the sample was again loaded in tension to assess the strength of crack repair. Comparisons were made between controls without chemicals and samples with chemicals for release. Old cracks did not reopen, but new ones formed. Quantified bend tests verified that release of adhesive from fibers does repair cracks, creates crack reopening resistance, and impedes crack propagation. Other samples were tested in impact. Visual assessment verified the propagation of matrix cracks through the repair fibers, the release of chemical from filled repair fibers into matrix cracks, and movement of the chemical along the length of the crack. After a cure time, it was observed that cracks filled with repair adhesives were rebonded. The migration of chemical through the entire crack length was observed. There was resistance to displacement at the crack location, that is, repair. Impact tests demonstrated that release of adhesive restored impact strength.

Design of the System

In general, materials capable of passive, smart self-repair consist of several parts: (1) an agent of internal deterioration, such as dynamic loading, that induces cracking, (2) a stimulus to release the repairing chemical, (3) a fiber or bead, (4) a coating or fiber wall that can be removed or changed in response to the stimulus, (5) a chemical carried inside the fiber, and (6) a method of hardening the chemical in the matrix in the case of cross-linking polymers or a method of drying the matrix in the case of a monomer.[3]

Several questions should be asked:

- Does the repair system cause any detriment to the mechanical properties of the product?
- What strength does the repaired product have? Is damage stopped? Is the matrix reinforced?
- Can it respond to impacts or earthquakes? Cyclic loading? Can it dampen?
- What is the speed or rate at which triggering and self-repair occurs?
- Does the self-healing member repair the types of damage and restore lost strength associated with the damage of the end use chosen?
- Will the self-repairing system survive all processing and environmental conditions to which the bridge or airplane, for example, will be subjected, including high temperatures, pressures, chemical spills?

After developing several examples of time release in smart materials, Dry,[3] who has a patent on time release in a variety of matrices, drew up the following list of attributes desirable for a successful design optimization of such a system for different applications.

1. The types of problems or distress that can be effectively treated by chemical release must be determined. The attributes of those types of distress are:
   - There must be a time-dependent problem of durability relating the exterior environment to the matrix conditions.
   - The problem must be important in cost and a frequent cause of deterioration.
   - The distress could be treatable by the chemical released.
2. The internal stimulating agent could be chosen if needed to allow the matrix to accept the chemical to be released, or provide other chemicals (for cross linking).
3. The release mechanism from the fiber should be designed for the specific application.
4. The types of chemicals to be released — that is, the encapsulant — must be effective in treating the specific type of distress.
5. The physical properties of the release agent or encapsulator — that is, material and shape — should be tailored to the treatment of the specific environmental distress, and to other needs of the design of the component. The method of encapsulating chemicals into the fiber or microcapsules must be an efficient and inexpensive process.
6. The stimuli or source of energy or change must be such that it can cause the chemical to be released from the fibers.
7. The matrix must be such that it can accept the encapsulant in form, chemistry, and volume; accept the chemical as it is released; and accept the stimuli action necessary to release the chemical.

Time Release of Repair Chemicals into Matrices

Flow in matrices can be described by D'Arcy's law for laminar flow through media. The treatment for cracks in which the chemicals are released internally only with the specific stimulus of loading is described by the flow of the chemical release from the fiber into the matrix. In the case of polymer matrices with little porosity, single or double cross-linking adhesive may be used. To apply D'Arcy's law, one must know the following:

- The force and frequency of the intrusion of the degrading agent, that is, loading.
- The condition of the matrix for ion content, permeability, at the time of intrusion.
- The force of the stimulus, which releases the chemical from the fiber.
- The factors affecting mixing of the cross-linking binary system of chemicals (if these are used).
- The temperature of the materials.
- The viscosity and density of the chemical released from the fiber.
- Diffusing into mechanical damage, that is, cracks, delaminations, debonds.
- The cross-section area of the specimen.
- The rate of curing of adhesive chemical.
- The affect of repair location and physical constraint on adhesive modulus of elasticity.

Self-Healing Composites and Glass

Controlled release of crack or filler material from a stretched, cracked, or debonded repair fiber is used to seal matrix microcracks and rebond damaged interfaces. Chemicals under investigation include adhesives and cross-linking or air-cured polymer/monomers. Two different time release designs are being investigated for the repair of microcracking in polymer matrix composites. The first is tensile or flexural loading breaking the hollow fiber, releasing its chemical. The second is tensile loading causing debonding of (hollow porous) fiber from matrix. Stripping away the coating releases its chemical through the pores of the fiber wall. Natural process design has shown both of these methods to be effective.

An investigation was made into the development of polymer matrix composites that have the ability to self-repair internal cracks due to mechanical loading.[1] It focused on the cracking of hollow repair fibers dispersed in a matrix and the subsequent release of repair chemicals, which results in the sealing of matrix cracks, the restoration of matrix strength, and the ability to retard crack propagation. It was found that cracking of the repair fiber and release of repair chemical could be achieved. This was verified by visual and auditory inspection. Figure 4 shows the release into a polymer matrix. Impact fracture and bend tests were performed and revealed the ability of this system to fill and repair cracks, restore strength, and retard crack growth. Fiber pullout tests were also done that revealed that the fibers were rebonded.

Laboratory Research on a Self-Repairing System

The research concerns repair of damage by incorporating time release repair fibers into a metal fiber–reinforced epoxy matrix composite. Bending samples measuring 1 × 1 × 4.5 in. had two reinforcing layers in the tension zone (lower 0.5 in. of the sample). Both layers contained three hollow glass fibers (with an average inner diameter of 0.8 mm) and three metal wires running the length of the samples. Both layers were below the midpoint of the samples section. Two different mechanisms for self-repair of polymer composites were investigated: release of single polymer and release of a cross-linking a two-part polymer. The single polymer used was cyanoacrylate; the two-part polymer used was a two-part epoxy. The design for time release that was investigated for the repair of cracking in polymer matrix composites is one in which loading breaks the hollow repair fibers, causing them to release their repair chemical. The release was photo documented under a microscope.

The purpose of bend tests was to evaluate the repair of cracks, the resistance to crack reopening, and the ability to retard further crack growth using both one- and two-part adhesives. Allowing eight months after the initial bend tests in which the glass fibers were broken and repair chemicals were released, to let the adhesive set up, the matrix bars were again tested in bending. These tests were to ascertain crack repair over time. In all but one of the four experimental samples the original crack did not reopen, but new ones opened. In the two control matrix bars without adhesive in the fibers, the old cracks reopened. They had not been repaired. These results show that release of adhesives from hollow glass fibers into matrix cracks is an effective way of repairing cracks, resisting further cracking at the original crack and allowing new cracks to form elsewhere in the matrix. Quantified bend tests verified that release of adhesive from fibers does repair cracks, create crack reopening resistance, and impede or stop crack growth (see Fig. 5). Quantified impact tests demonstrated that internal delayed release of adhesive restores the impact strength and ability to deflect while carrying a load.[3]

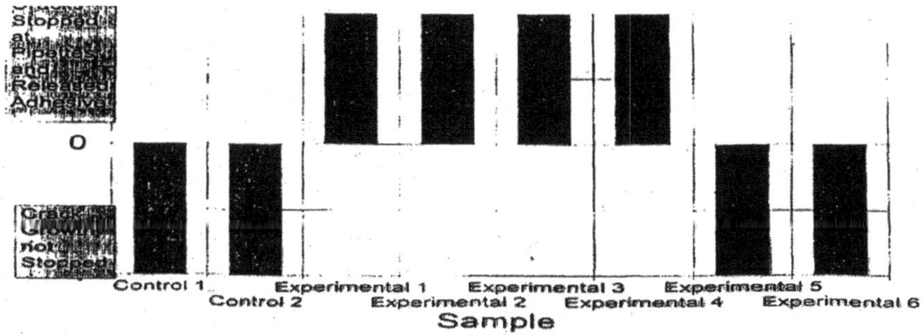

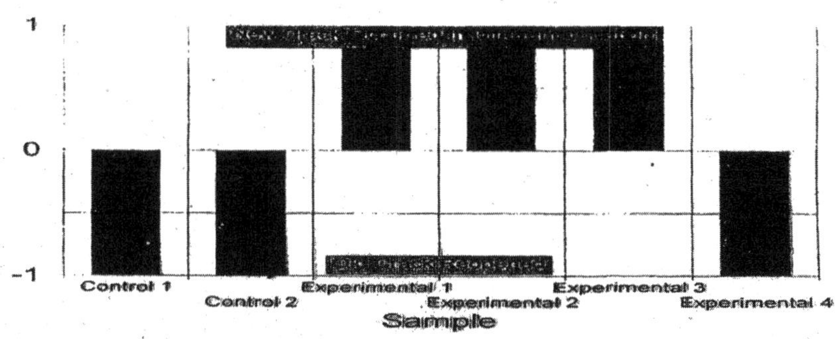

Figure 5. Results from the first and second bend tests on samples such as those at the right. In the first test, the number given indicates number of layers of pipettes (of the three in each sample) that cracked in the bend test. In second test, it is noted whether or not the original cracks reopened.[3]

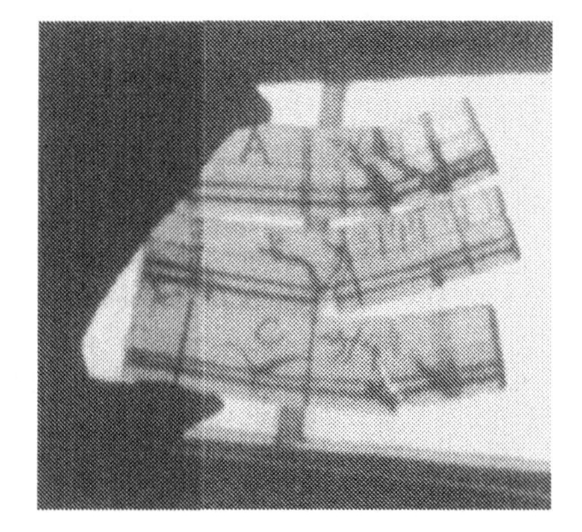

The research on self-repairing glass involves the placement of hollow adhesive-filled fibers into the glass matrix itself. Impact would release the repair chemical into the cracks. In the glass samples already tested (Fig. 6), the filled fibers were between glass plates. Upon impact, the glass pates broke and the filled fibers broke and released the dyed adhesive onto the

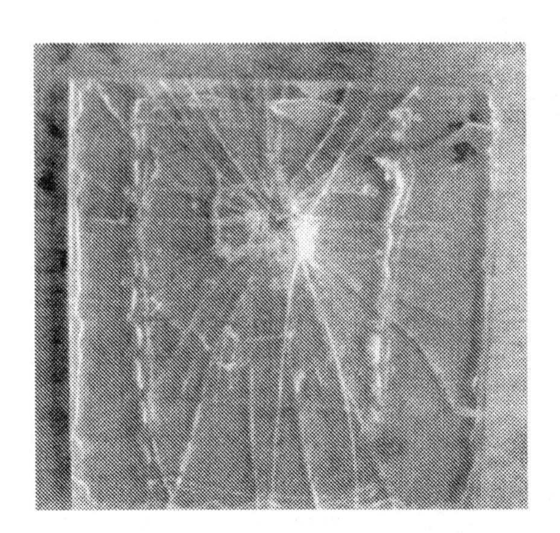

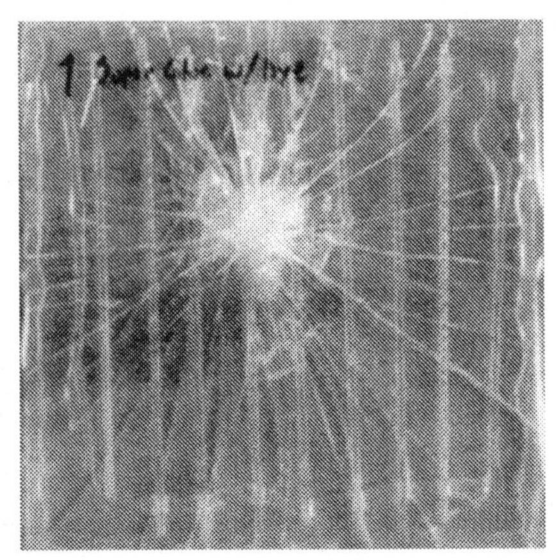

Figure 6. Two samples of double glass plates sandwiching dyed, adhesive-filled, hollow glass fibers in between.

surface of the plates. Some repair was accomplished but it is anticipated that much more successful repair will occur when the repair fibers are in the same plane as the glass. This is based on our experience with polymer repair.

References

1. C. M. Dry and N. Sottos, "Passive Smart Self-Repair in Polymer Matrix Composite Materials"; in *Proceedings of the 1993 North American Conference on Smart Structures and Materials,* Vol. 1588, SPIE, 1–4 February 1993.
2. C. M. Dry and William McMillan, "Three-Part Methylmethacrylate Adhesion System as an Internal Delivery System for Smart Responsive Concretes," *Smart Mater. Struct.,* (1996), pp. 297–300.
3. C. Dry, "Procedures Developed for Self-Repair of Polymer Matrix Composite Materials," *Composite Struct.,* **35**, 263–269 (1996).

9 780470 051467